PRIMARY HEALTH C.
PEOPLE, PRACTICE, P

T0228383

# Geographies of Health

*Series Editors*

Allison Williams, Associate Professor, School of Geography and Earth
Sciences, McMaster University, Canada
Susan Elliott, Dean of the Faculty of Social Sciences,
McMaster University, Canada

There is growing interest in the geographies of health and a continued interest in what has more traditionally been labeled medical geography. The traditional focus of 'medical geography' on areas such as disease ecology, health service provision and disease mapping (all of which continue to reflect a mainly quantitative approach to inquiry) has evolved to a focus on a broader, theoretically informed epistemology of health geographies in an expanded international reach. As a result, we now find this subdiscipline characterized by a strongly theoretically-informed research agenda, embracing a range of methods (quantitative; qualitative and the integration of the two) of inquiry concerned with questions of: risk; representation and meaning; inequality and power; culture and difference, among others. Health mapping and modeling, has simultaneously been strengthened by the technical advances made in multilevel modeling, advanced spatial analytic methods and GIS, while further engaging in questions related to health inequalities, population health and environmental degradation.

This series publishes superior quality research monographs and edited collections representing contemporary applications in the field; this encompasses original research as well as advances in methods, techniques and theories. The *Geographies of Health* series will capture the interest of a broad body of scholars, within the social sciences, the health sciences and beyond.

# Primary Health Care:
# People, Practice, Place

*Edited by*

VALORIE A. CROOKS
*Simon Fraser University, Canada*

GAVIN J. ANDREWS
*McMaster University, Canada*

LONDON AND NEW YORK

First published 2009 by Ashgate Publishing

2 Park Square, Milton Park, Abingdon, Oxon OX14 4RN
711 Third Avenue, New York, NY 10017, USA

*Routledge is an imprint of the Taylor & Francis Group, an informa business*

First issued in paperback 2016

**British Library Cataloguing in Publication Data**
Primary health care : people, practice, place. - (Ashgate's
    geographies of health series)
    1. Primary health care 2. Medical geography
    I. Crooks, Valorie A. II. Andrews, Gavin J., 1970-
    362.1

**Library of Congress Cataloging-in-Publication Data**
Primary health care : people, practice, place / [edited] by Valorie A. Crooks and Gavin J. Andrews.
        p. cm. -- (Ashgate's geographies of health series)
    Includes bibliographical references and index.
    ISBN 978-0-7546-7247-0
    1. Medical geography. 2. Primary care (Medicine) I. Crooks, Valorie A.,
    1976- II. Andrews, Gavin J., 1970- III. Series.
    [DNLM: 1. Primary Health Care. 2. Geography. 3. Health Services Accessibility. 4. Healthcare Disparities. 5. Social Environment. W 84.6 P95234 2008]
    RA792.P75 2008
    614.4'2--dc22

                                                                                2008038857


ISBN 13: 978-0-7546-7247-0 (hbk)
ISBN 13: 978-1-138-26243-0 (pbk)

# Contents

# List of Figures

# List of Tables

# Notes on Contributors

**Gina Agarwal** was trained as a doctor, then a GP in England, and as an academic in Canada. She is now an Assistant Professor in the Department of Family Medicine, McMaster University (Canada). Gina works as a family physician and teaches at the academic McMaster Family Practice. Gina's main areas of interest are primary care diabetes and its epidemiology, chronic illness and the doctor-patient relationship, and continuity of primary care.

**Gavin J. Andrews** is Professor and Chair, Department of Health, Aging and Society at McMaster University (Canada). His geographical research investigates a broad range of subjects including age and aging, sport and fitness, complementary medicine and the health histories of places. Most relevant to the current book however is his interest in health care organizations, workers and work. To date, Gavin has published over 80 journal articles and contributions to collections. He is the North American Editor for the *International Journal of Older People Nursing*, and Associate Editor *Journal of Applied Gerontology.*

**Pauline Barnett** is an Associate Professor in the Department of Public Health and General Practice at the Christchurch campus of the University of Otago (New Zealand). Her primary research interest has been in public sector and health restructuring, particularly primary care and mental health. Her recent research has focused on health sector governance and the reform of primary care. She is a member of the National Health Committee which advises the New Zealand Minister of Health and is a member of several community boards and advisory committees for mental health, disability and primary care agencies. In 2004 she was designated 'Public Health Champion' by the Public Health Association of New Zealand.

**Ross Barnett** is Professor of Geography at the University of Canterbury, Christchurch, New Zealand. His research has traditionally focused on the geography of health services and health sector restructuring, but in recent years has shifted to examining ethnic and social inequalities in health with particular reference to smoking cessation, melanoma, cancer survival and diabetes. In addition, he has continued his work in the geography of health services particularly in primary care, the voluntary sector and practice variations in hospital admissions. He is Chair of the Social Science Research Centre Board at the University of Canterbury and a member of Housing New Zealand's Evaluation Advisory Group overseeing the Healthy Housing Programme. Reflecting his interest in health he is also a keen orienteer!

**Ivy Bourgeault** is an Associate Professor of Sociology and Health, Aging and Society and Canada Research Chair in Comparative Health Labour Policy at McMaster University (Canada). She has published in national and international journals and edited volumes on complementary and alternative medicine, primary care and health professions and health policy.

**Sandra Chen** is an Advanced Practice Nurse in Mental Health. She received her Bachelor of Science in Nursing and her Master's of Nursing at the University of Toronto. She is currently the Clinical Practice Leader for the Mental Health Program of Rouge Valley Health System in Ajax, Ontario, Canada. Her areas of interest include child and adolescent mental health, participatory action research, mental health promotion, and community development.

**David Conradson** is Senior Lecturer in Human Geography at the University of Southampton, UK. His research interests include the dynamics of voluntary welfare provision, organizational spacings of subjectivity, and psychosocial wellbeing. His publications in these areas include work on community-drop in centres, therapeutic environments and faith-based welfare services. Current projects are examining shifting landscapes of residential aged care provision in New Zealand and popular engagement with places of retreat in England.

**Valorie A. Crooks** is an Assistant Professor in the Department of Geography at Simon Fraser University in British Columbia, Canada. She completed her PhD in 2005 at McMaster University (Hamilton, Ontario) and in 2005–06 undertook a postdoctoral fellowship at York University (Toronto, Ontario). Much of her research has focused on women's experiences of chronic illness, some of which is centered on exploring how power is enacted in place through the doctor-patient interaction and how such performances are constrained and shaped by larger forces including the organization and delivery of (primary) health care. She has also recently been involved in several projects related to continuity of care in family practice and another pertaining to female newcomers' access to primary mental health care. Finally, she has long-standing interests in the geographies of disability and health-related social policies/programs and emerging interests in rural palliative care and family caregiving.

**Leah Gold** obtained her Master's in geography at Simon Fraser University, British Columbia, Canada in 2006. She completed a thesis entitled *Negotiating Health and Identity: Peruvian Villagers' Discourses on the Uneasy Co-existence of Primary Health Care and Indigenous Medicine*, examining the interstices between the differing ideologies of indigenous medicine and western medicine. Her broader research interests include cultural factors influencing the negotiation of health care choices, social relations that are (re)produced through the use of local indigenous or alternative medicine, and biomedicine as a representation of western cultural. She currently lives in Århus, Denmark.

**Neil Hanlon** is an Associate Professor of Geography at the University of Northern British Columbia (Canada). His research interests include such topics as the restructuring of acute care, the regionalization of health governance, access and utilization issues in rural and remote places, community level responses to population aging, and the intersections of formal and informal sector systems of support in a variety of Canadian settings. His work has published in leading journals such as *Social Science & Medicine, Health & Place, Health & Social Care in the Community* and *Health Policy.*

**Daniel Hollenberg** received his PhD in Community Health from the University of Toronto. He is currently a CHSRF postdoctoral research fellow in the Department of Health, Aging and Society at McMaster University and at the University of Toronto (Canada), Centre for International Health, where he is examining the introduction of CAM into Canadian hospitals. Combining both critical anthropology and sociology theory and methods, Daniel's research focuses on the biomedical and CAM professions, and the sociocultural push to blend them into 'integrative' medicine. His additional research interests include the role of traditional medicine in primary health care and cross-cultural studies of healing systems.

**Robin Kearns** is Professor of Geography at the University of Auckland (New Zealand). His research focuses on the role of place in shaping health experience and the place of (particularly health care) services in geographies of everyday life. Specifically, he has investigated the effects of social and economic restructuring on community well-being, changing geographies of mental health care, and children's experience of neighbourhoods. Robin's current funded research includes: neighbourhoods and physical activity; activism in the voluntary sector; maintaining home while aging in place; and changing meanings of coastal environments. He obtained his PhD at McMaster University (Canada) in 1987 supported by a Commonwealth Scholarship. He is author of numerous refereed papers and book chapters as well as two books with Wilbert Gesler, the most recent being *Culture/Place/Health* (Routledge, 2002). He is Associate Editor of two journals: *Health and Social Care in the Community* and *Heath & Place* and is a ministerial appointment on New Zealand's National Health Committee.

**Jennifer Lapum** is an Associate Professor in the Daphne Cockwell School of Nursing at Ryerson University (Ontario, Canada). She is also a PhD candidate at the Lawrence S. Bloomberg Faculty of Nursing, University of Toronto. She is a registered nurse with over ten years of clinical experience pertaining to the intensive care unit and cardiac surgery. She has expertise in qualitative approaches, particularly narrative inquiry. Her work has been published in *Qualitative Inquiry,* the *International Journal of Qualitative Methods, Social Science & Medicine,* the *Canadian Medical Association Journal,* and the *Canadian Journal of Cardiovascular Nursing.*

**Doris Leung** is a PhD candidate at the Lawrence S. Bloomberg Faculty of Nursing at the University of Toronto (Canada). Her studies focus on existential issues related to palliative and supportive care in cancer nursing. Having practiced as a Registered Mental Health Nurse and Clinical Educator for more than fourteen years, she not only draws upon her education, but also on professional and personal experiences to inform her writing about health care. She has published peer-reviewed abstracts and articles in the areas of mental health, psycho-oncology, and qualitative research, one of which was recently in the *International Journal of Palliative Care Nursing*.

**Graham Moon** is Professor of Geography at the University of Southampton, UK. He has published over 100 papers on various themes in health geography and attracted funding from research councils, industry and governments. He edits the journal *Health & Place*. His current research is motivated by a concern to identify the difference place makes to health through innovative research methods of both a quantitative and a qualitative nature, and is grounded in empirical studies from the UK, other European settings, Canada and New Zealand. He focuses on two themes: the secondary analysis of large-scale spatially-referenced datasets concerning health-related behaviour and health service use; and the spatial implications of health service reform. In the latter area his concern is with the changing geographical organization of health services and with discourses of place and space in health care policy. He has particular interests in primary care provision, especially in relation to notions of 'community.'

**Pat Neuwelt** is a Senior Lecturer in public health at the School of Population Health, the University of Auckland (New Zealand). Pat worked for ten years as a general practitioner with disadvantaged communities in urban and rural settings, in both Canada and New Zealand. She is now a public health medicine specialist, engaged in research on comprehensive primary health care as the interface between public health and primary care. Pat gained her medical degree from McMaster University and has postgraduate qualifications in family medicine (Memorial University, Canada), palliative care (Flinders University, Australia) and public health (the University of Auckland and University of Otago, New Zealand). Pat's doctoral research focused on community participation in primary health care in New Zealand, out of which she published a book entitled *Community Participation Toolkit: Resource for Primary Health Organisations* (Steele Roberts 2007). She has published on a number of public health topics, including access to primary care.

**Jessica Peterson** is a PhD candidate at the Lawrence S. Bloomberg Faculty of Nursing, University of Toronto (Canada). After graduating with a bachelor's degree in nursing, Jessica worked as a staff nurse on medical-surgical units in the US and Canada specializing in the post-operative care of patients following joint replacement. The goal of her dissertation research is to identify determinants of job satisfaction and well-being for newly graduated nurses. Additionally she is completing a diploma in health services and policy research through the Ontario Training Centre.

Professor **Mark W. Rosenberg** received his PhD (1980) from the London School of Economics (LSE) and Political Science. He is currently a Professor of Geography and cross appointed as a Professor in the Department of Community Health and Epidemiology at Queen's University (Canada).

**Nadine Schuurman** is a GIScience (GIS) researcher with a focus on improving population and health services data and analysis. She is an Associate Professor in the Department of Geography at Simon Fraser University (Canada) and holds a New Investigator Award from the Canadian Institutes of Health Research and a Michael Smith Foundation for Health Research Scholar Award. Her GIS laboratory is dedicated to the development of *new technologies, methodologies and protocols* for studying population health and health services. Her goal is to assist policy makers and health administrators understand and rationalize choices about service provision based on spatial data, population vulnerability, access, service availability, trajectories and outcomes.

**Mark W. Skinner** is Assistant Professor of Geography at Trent University, Canada. Mark's research interests are in health, rural and social geography, and his current work focuses on the interdependence of change among the voluntary sector, health and social care, and ageing rural communities. He is involved in several projects and collaborations on the evolving role of voluntarism in rural parts of Canada, France and New Zealand, with funding from various agencies including national research councils and community-based organizations. His work appears in edited collections published by Cambridge University Press, Elsevier and Policy Press, and in refereed journals such as *Environment and Planning C: Government and Policy, Health & Place* and *Social Science & Medicine.*

**Janine L. Wiles** received her PhD (2001) from Geography at Queen's University. She is currently a Senior Lecturer in Social and Community Health in the School of Population Health at the University of Auckland (New Zealand). Her research interests are in geographies of health and health care and social gerontology, including health care restructuring, older people and their relationships to their social and physical environments, family or 'lay' caregiving to older people at home, community-based care, home as a site for care, geographical gerontology, gender, and social difference.

**Nicole M. Yantzi** received her PhD (2005) from the Department of Geography, Queen's University. She is currently an Assistant Professor of Geography and Faculty Investigator at the Centre for Rural and Northern Health Research at Laurentian University, Sudbury, Canada. Nicole's research intersects the geographies of children and youth, disability, and health and health care. Her current research examines how the socio-spatial exclusion and marginalization of children with disabilities in the key spaces of childhood (homes, schools/playgrounds and neighbourhoods/playgrounds) impacts their ability to fully participate in their communities. She is especially interested in the challenges that those living in rural and northern areas negotiate in developing inclusive play spaces. Her work appears in referred journals such as *Gender, Place and Culture, Health and Social Care in the Community* and *Social Science & Medicine*.

# Chapter 1

# Thinking Geographically About Primary Health Care

Valorie A. Crooks and Gavin J. Andrews

The overriding goal of the geographies of health and health care is to 'construct accounts of why place matters' to these fields (Mohan 2000, 330). Scholarship in this vibrant area of the discipline of human geography has had a significant impact on numerous areas of health and health care inquiry more broadly, including our conceptualization of the social determinants of health and also issues related to the organization and delivery of health care. However, as Gesler and Kearns (2002, 139) point out: '[u]ntil the mid-1990s, geographical analysis of health care ... was often undertaken by examining patterns of service utilization. Furthermore, the sites of service provision were viewed as locations rather than as contributors to, and constituents of, health care landscapes.' More recently there has been a flourishing of innovative research undertaken by geographers and others that offers unique insights into how we conceptualize the geographies of health care in a much broader sense (see Chapter 2), some of which specifically connects to primary health care (PHC). In this book we showcase this work through interrogating the landscapes that inform the very nature and principles of PHC and ultimately shape how this care is both delivered and received.

In this chapter we provide an important background for the book and also an overview of what's to come in the remaining chapters. We first engage in considering the very nature of PHC. In doing so we begin with the 1978 Declaration of Alma-Ata and continue up to the present time where PHC reform and renewal are happening. We then move to introduce the geographies of PHC. It is observed that while geographers have a long-standing interest in this topic, few studies explicitly or fundamentally engage PHC as a concept. Through focusing on access, equity, and community/participation – central elements of the PHC approach – we frame this area of health geography inquiry. In the final section we provide an overview of the book and summaries of the contributed chapters.

## What is Primary Health Care?

In attempting to understand what PHC is, one must start with the Declaration of Alma-Ata set out by the World Health Organization (WHO) in 1978. This Declaration is the outcome of a week-long conference hosted by the WHO in

Alma-Ata, USSR focused specifically on PHC. Over 3,000 delegates representing international bodies, such as the United Nations Children's Fund (UNICEF), and national governments attended the conference (Cueto 2004). PHC was viewed as a way to achieve 'health for all' which was a significant goal of the WHO at the time[1] (Pappas and Moss 2001). We must note, however, that using the Declaration as a starting point does not mean that the kind of care central to the principles of PHC outlined within it was not delivered or envisioned prior to this point (see also Cueto 2004; WHO-SEA 2006). Rather, the Declaration marks the point at which PHC was formalized as a form of care and also defined at a global scale. The Declaration is detailed in Table 1.1.

Both the Declaration and the conference in Alma-Ata remain important moments in the development of PHC as an approach to care provision. Werner (1995) points out that the Declaration is a 'landmark' because it identifies health and access to health care as basic human rights. Further, its emphasis on community and community members is what set the PHC approach apart from other health service delivery models (Litsios 2004). Cueto (2004) has identified three important ideas shared in the Declaration which have been widely cited: (1) use of appropriate technology; (2) opposition to medical elitism; and (3) health as a tool for social development. These ideas and others central to the Declaration were built into the WHO's vision of PHC for specific reasons. At the time of the conference in Alma-Ata there was an increasing desire not to transplant Western models of care to lesser developed nations, an emergence of the determinants of health approach, and an increased recognition of the effectiveness of grassroots and/or local approaches to health care (Cueto 2004; Werner 1995). Successful examples of this community-based approach to care in China, Tanzania, Sudan, and Venezuela in the 1960s and 70s created a general enthusiasm within the WHO about PHC. For example, the 'barefoot doctor movement' in China at the time is often cited as providing a significant impetus to the formalization of PHC. These factors led PHC to be viewed as a way forward in achieving 'health for all' and overcoming health disparities within and between nations at the time through the provision of equitable and accessible care (i.e., medical treatment, preventative care, health promotion, some forms of social care) in, by, and for communities.

---

1   The WHO's goal of achieving health for all remains. At the time of the conference in Alma-Ata it was declared that this goal would be reached by the end of the twentieth century. By the early 1990s it became clear that this would not be the case. By 1995 the WHO had declared this to be a twenty-first century goal (Pappas and Moss 2001).

**Table 1.1 The Declaration of Alma-Ata**

I.   The Conference strongly reaffirms that health, which is a state of complete physical, mental and social wellbeing, and not merely the absence of disease or infirmity, is a fundamental human right and that the attainment of the highest possible level of health is a most important world-wide social goal whose realization requires the action of many other social and economic sectors in addition to the health sector.

II.  The existing gross inequality in the health status of the people particularly between developed and developing countries as well as within countries is politically, socially and economically unacceptable and is, therefore, of common concern to all countries.

III. Economic and social development, based on a New International Economic Order, is of basic importance to the fullest attainment of health for all and to the reduction of the gap between the health status of the developing and developed countries. The promotion and protection of the health of the people is essential to sustained economic and social development and contributes to a better quality of life and to world peace.

IV.  The people have the right and duty to participate individually and collectively in the planning and implementation of their health care.

V.   Governments have a responsibility for the health of their people which can be fulfilled only by the provision of adequate health and social measures. A main social target of governments, international organizations and the whole world community in the coming decades should be the attainment by all peoples of the world by the year 2000 of a level of health that will permit them to lead a socially and economically productive life. Primary health care is the key to attaining this target as part of development in the spirit of social justice.

VI.  Primary health care is essential health care based on practical, scientifically sound and socially acceptable methods and technology made universally accessible to individuals and families in the community through their full participation and at a cost that the community and country can afford to maintain at every stage of their development in the spirit of self-reliance and self-determination. It forms an integral part both of the country's health system, of which it is the central function and main focus, and of the overall social and economic development of the community. It is the first level of contact of individuals, the family and community with the national health system bringing health care as close as possible to where people live and work, and constitutes the first element of a continuing health care process.

VII. Primary health care:

    a. reflects and evolves from the economic conditions and sociocultural and political characteristics of the country and its communities and is based on the application of the relevant results of social, biomedical and health services research and public health experience;

    b. addresses the main health problems in the community, providing promotive, preventive, curative and rehabilitative services accordingly;

    c. includes at least: education concerning prevailing health problems and the methods of preventing and controlling them; promotion of food supply and proper nutrition; an adequate supply of safe water and basic sanitation; maternal and child health care, including family planning; immunization against the major infectious diseases; prevention and control of locally endemic diseases; appropriate treatment of common diseases and injuries; and provision of essential drugs;

**Table 1.1 continued**

d. involves, in addition to the health sector, all related sectors and aspects of national and community development, in particular agriculture, animal husbandry, food, industry, education, housing, public works, communications and other sectors; and demands the coordinated efforts of all those sectors;

e. requires and promotes maximum community and individual self-reliance and participation in the planning, organization, operation and control of primary health care, making fullest use of local, national and other available resources; and to this end develops through appropriate education the ability of communities to participate;

f. should be sustained by integrated, functional and mutually supportive referral systems, leading to the progressive improvement of comprehensive health care for all, and giving priority to those most in need;

g. relies, at local and referral levels, on health workers, including physicians, nurses, midwives, auxiliaries and community workers as applicable, as well as traditional practitioners as needed, suitably trained socially and technically to work as a health team and to respond to the expressed health needs of the community.

VIII. All governments should formulate national policies, strategies and plans of action to launch and sustain primary health care as part of a comprehensive national health system and in coordination with other sectors. To this end, it will be necessary to exercise political will, to mobilize the country's resources and to use available external resources rationally.

IX. All countries should cooperate in a spirit of partnership and service to ensure primary health care for all people since the attainment of health by people in any one country directly concerns and benefits every other country. In this context the joint WHO/UNICEF report on primary health care constitutes a solid basis for the further development and operation of primary health care throughout the world.

X. An acceptable level of health for all the people of the world by the year 2000 can be attained through a fuller and better use of the world's resources, a considerable part of which is now spent on armaments and military conflicts. A genuine policy of independence, peace, détente and disarmament could and should release additional resources that could well be devoted to peaceful aims and in particular to the acceleration of social and economic development of which primary health care, as an essential part, should be allotted its proper share.

*Source*: WHO 1978, n.p.

Early versions of this Declaration were shared throughout the WHO and its respective regional offices for up to a year before the conference in Alma-Ata happened. Tejada de Rivero (2003), the former Deputy Director-General of the WHO, notes that the Declaration went through no less than 18 drafts and circulation at six regional WHO meetings prior to the conference. This resulted in much of the final wording of the Declaration looking nothing like what it was originally drafted to convey (Litsios 2002). As Tejada de Rivero (2003, n.p.) recounted: 'The draft that was officially presented [at the conference] contained a few changes that, in hindsight, contributed to the later distortion of the original concepts. Many delegations and individual delegates fought to include details that had more to do with medical specialties than with health.' Further to this, the conference was proposed in 1974 (Litsios 2002), giving some national representatives and WHO employees up to four years to digest the PHC concept and prepare their responses. These details are quite significant because *immediately* after the Alma-Ata conference critiques of the WHO's vision of PHC were circulated, many of which had taken shape in advance of any formal release of the Declaration.

The Alma-Ata Declaration was immediately criticized as being too broad in scope, thereby making the WHO's vision of PHC difficult to implement. An outcome of this was that there was a selective interpretation and implementation of the principles of PHC that were set out by the WHO (Cueto 2004; Haines et al. 2007). This has come to be known as 'selective PHC' and typically involves a 'package of low-cost technical interventions to tackle the main disease problems of poor countries' (Cueto 2004, 1868). Selective PHC often involves temporally and spatially limited initiatives to counter specific health issues, often characterized by GOBI-style initiatives. GOBI is an acronym for a set of four specific interventions (growth monitoring, oral rehydration, breast feeding, and immunization) that were championed by UNICEF and others – and to a certain extent continue to be – as a useful model of selective PHC (Wisner 1988). Such interventions can be funded by external aid organizations and implemented in nations with little or no existing health care infrastructure. While these interventions may assist in a limited way with achieving 'health for all', they do not hold true to the basic tenants of PHC put forth in the WHO's Declaration including that care is delivered by, for, and in communities. Criticisms of selective PHC are numerous and point out that it focuses only on basic health services (Cueto 2004), that it takes control and power away from developing nations in particular and places it with international funding bodies (Hall and Taylor 2003) and non-governmental organizations (Litsios 2002), and that this approach does not adequately recognize the importance of factoring local contexts (e.g., political involvement, cultural norms) into health care initiatives (Walsh 1988). That UNICEF championed a selective interpretation of and approach to PHC has been particularly criticized given that it was a signatory to the (non-binding) Alma-Ata Declaration and a sponsor of the conference. Though, it has been argued that (Taylor and Jolly 1988, 972):

The concept of 'selective primary health care' [SPHC] was built into the original definition of PHC in the background document for the Alma-Ata conference. This is important to underline, since this point has often been misinterpreted, especially by proponents of SPHC arguing that CPHC [comprehensive PHC] ignored the need for explicit priorities.

The subsequent undertaking of selective PHC initiatives following the Alma-Ata Declaration has been cited as a major failure of the WHO's vision of PHC and its ability to put it into action. There are, however, numerous other lasting critiques and challenges to implementing PHC. Cueto (2004) points out that some medical professionals were (and are) resistant to the PHC approach because of concerns over loss of prestige and power through increasing the involvement of non-professionals in delivering care in the community. At the same time, Haines et al. (2007) point out that the increased reliance on volunteer labour by community members makes PHC difficult to sustain. While the focus of the Declaration is very much on the community and traditional practices, Wakai (1995) notes that it places traditional medicine secondary to Western hi-tech medicine, despite its predominance in much of the developing world, provision of first contact care, and clear potential to connect people to orthodox medicine. Furthermore, Tejada de Rivero (2003) contends that misunderstandings resulting from both the Declaration and conference at Alma-Ata have significantly challenged the implementation of the WHO's vision of PHC. Among these misunderstandings were the questionable word choices made when translating the Declaration into other languages and the oversimplification of reported successes in developing nations. Other cited obstacles to implementing PHC include (Hall and Taylor 2003; Wakai 1995):

1. the structural adjustment programs in place in some developing nations;
2. the World Bank takeover of health care policy making in many developing nations;
3. civil wars, natural disasters, and major health crises (e.g., HIV) that constrained health care development in sub-Saharan nations in particular;
4. a lack of long-term political commitment to PHC; and
5. difficulties finding funding to back comprehensive PHC initiatives in developing nations.

These obstacles often justified, if not necessitated, the selective implementation of PHC in developing nations.

While response to the Declaration and specifically the PHC approach in developing nations was very much to implement selective models of care, developed nations had a much different reaction. That said, it is important to note that developed nations very much pushed for the selective implementation of PHC in developing nations in order to maintain a certain degree of power and control over how health care monies were spent in these places (Hall and Taylor 2003; Werner 1995). In countries such as Canada, New Zealand, and the United Kingdom

(UK) – those which are focused on the most in this book – health care systems were already in place prior to the Declaration being made. These governments and their health care decision-makers struggled with what it meant to offer PHC. Numerous interpretations of PHC were made in response to the WHO's definition, and continue to this day due to differences in national health care systems (Haines et al. 2007). In general, a shift in the focus of care away from the hospital and into the community was and is viewed as in keeping with the PHC approach. PHC is often interpreted to mean care that is universal, comprehensive, affordable, and delivered equitably (Hall and Taylor 2003). A focus on 'first-contact care' was viewed to be at the core of PHC by nations with existing health care systems. More specifically, putting in place or ensuring equitable and affordable first-contact care delivered via existing primary (medical) care systems was where the PHC approach resonated most strongly and sensibly for developed nations (Haines et al. 2007).

Primary care systems are certainly implicated in the model of PHC promoted by the WHO. Primary care systems facilitate the provision of first-contact, comprehensive, and coordinated medical care (Cardarelli and Chiapa 2007) and are characterized by organization, funding, and delivery (Hutchison et al. 2001). These systems are reliant on family doctors or general practitioners (GPs) for medical care provision along with nurses and some allied health professionals. Primary care serves as the entry point to the formal health care system in many nations and is where core medical and preventative care are delivered (Schoen et al. 2004). From this brief overview of primary care systems they can be understood to be where first-contact medical care is delivered from in nations that have health care systems that are organized by primary, secondary, and tertiary care levels. A position statement from the Canadian Nurses Association helps to differentiate primary care from PHC (CNA 2005, 1):

> PHC includes basic medical and curative care at the first level (commonly referred to as primary care). PHC is also relevant to secondary and tertiary care. The PHC approach focuses on promoting health and preventing illness. The PHC approach means being attentive to and addressing the many factors in the social, economic and physical environments that affect health – from diet, income and schooling, to relationships, housing, workplaces, culture and environmental quality. In addition, the PHC approach places citizens and patients on an equal footing with health care professionals with respect to decision-making about health issues.

Because of a shared focus on first-contact care it is easy to understand why many nations with existing primary care systems have applied the principles of PHC most easily to this level of care. However, this interpretation of PHC embraces only a limited conceptualization of the vision laid out at Alma-Ata. Specifically, the application of the PHC approach to primary care systems tends only to reference principle VI of the Declaration and often misses out on an integrated understanding of community participation, among other key concepts. Because of this, a primary care system alone cannot fully effectuate PHC given its focus specifically on treatment and diagnosis.

Confusion regarding the similarities and differences between primary care and PHC and conflation of one with the other is a long-standing issue. Green (1987), for example, critically examined the uptake of the principles of the Declaration within the National Health Service (NHS) of the UK within the years following the conference at Alma-Ata. He found a general lack of awareness about PHC within the NHS and attributes this to three main factors: (1) confusion regarding the difference between PHC and primary (medical) care, in that much research framed as PHC is in fact about primary care; (2) the view that PHC is a policy strategy for developing countries and not applicable in developed nations; and (3) defensiveness within the NHS due to a slowing of resources and reorganization. These factors clearly serve as significant barriers and provide a clear illustration about both the way in which a developed nation and its health care system may interpret the Declaration and its principles and also the problem of a general inability to differentiate primary care from PHC.

In order to overcome the challenges of interpreting and applying the very broad model of PHC envisioned by the WHO in the Declaration, or of too narrowly applying it to primary care alone or in a selective fashion, many governments, health councils, health professional bodies, and the like have created their own working definitions of PHC. These definitions often reference the Declaration, and most often principle VI. Such definitions frame PHC in numerous ways, including as a paradigm (Cueto 2005), a policy (WHO-EM 2003), a philosophy (WHO-SEA 2006), an approach (CNA 2005), a model, or even a system of care. Each of these framings has a different meaning and ultimately different implications for the actualization of PHC. In order to avoid discussing this in the abstract in Table 1.2 we have summarized a range of framings (or even frameworks) for PHC in a single national context by using the case of Canada. From this table it is possible to understand the range of ways in which PHC is both interpreted and applied in specific contexts. To further exemplify this issue, in Table 1.3 we have summarized a limited number of definitions of PHC across provincial/territorial jurisdictions within Canada. In looking at this table it can be seen that some provinces/territories more closely align their definitions of PHC with the primary care system while others, most notably the Northwest Territories, reference the spirit and principles of the WHO's Declaration. These Canadian examples serve to illustrate the lack of international or even national consensus regarding how to define, interpret, and apply PHC both within and beyond health care systems.

**Table 1.2 Multiple framings of primary health care in Canada**

| Framing | Example |
|---|---|
| A system | PHC is 'community-based health professionals and programs that are the first point of contact with the health care system' (Health Council of Canada 2008, 6). |
| A set of qualities | Responsive, comprehensive, continuity, interpersonal communication, and technical effectiveness (Broemeling et al. 2006). |
| A range of services | Emphasizing health promotion, chronic illness management, and integration of services within a continuum of care (Health Council of Canada 2008). Also includes basic emergency care, referrals, primary mental health care, palliative care, healthy child development, primary maternity care, and rehabilitation services (Health Canada 2006). |
| An approach | PHC 'refers to an approach to health and a spectrum of services beyond the traditional health care system. It includes all services that play a part in health, such as income, housing, education, and environment. Primary care is the element within primary health care that focuses on health care services, including health promotion, illness and injury prevention, and the diagnosis and treatment of illness and injury' (Health Canada 2006, n.p.). |
| A sector | 'The PHC sector contributes to health system equity directly through its responsibility for distribution of care within this sector, and indirectly through control over prescriptions, referrals and hospital admissions' (Watson et al. 2005, 103). |
| A set of system components | Specific components of PHC are: (1) improved continuity and coordination; (2) early detection and action; (3) better information on needs and outcomes; and (4) incentives to support the adoption of new health care approaches (Health Council of Canada 2005). |
| A group of providers | Physicians, dieticians, home care workers, nurses, occupational therapists, physiotherapists, pharmacists, social workers, and other health care providers (Watson et al. 2005). |

*Source*: The author.

**Table 1.3  Primary health care across selected Canadian provinces and territories**

| Province/ Territory | Definition |
| --- | --- |
| British Columbia | The first and most frequent point of contact with the health care system. Whether it is a visit to the family doctor or from a home care worker, a trip to the pharmacist, mental health counsellor or school nurse, PHC is where new health problems are addressed, and where patients and providers work together to manage ongoing problems. The goal of PHC is to keep people healthier, longer, by preventing serious illness and injury through education and timely treatment of short-term or episodic problems. It also works to help patients manage chronic health illnesses appropriately, so they don't develop unnecessarily into medical crises. |
| New Brunswick | First contact with the health system by patients. Is provided by clinicians who are responsible for addressing a wide array of health care needs and developing a long-term relationship with their patients. Changes to delivery of PHC will result in better, faster access to services, in the community and throughout the health system. |
| Northwest Territories | The term 'primary health care' is used interchangeably with the term 'primary community care' to reflect the health and social services environment. It is the first level of care and usually is the first point of contact clients have with the health and social services system - that is, in partnership with the client, services are mobilized and coordinated in response to client needs to promote wellness, prevent trauma and illness, build capacity, provide support and care for community health and social issues and manage ongoing problems to sustain functional independent at an optimal level. |
| Ontario | The foundation of the health care system with a sustainable, long-term relationship between the inter-disciplinary health care teams and patient. It refers to the first level of care and the initial point of contact that a patient has with the health system. |
| Prince Edward Island | It is a philosophy and an approach to health care based on the principles of accessibility, public participation, health promotion, illness prevention, appropriate technology and inter-sectoral collaboration. |

*Source*: Health Council of Canada 2005, 24–25.

Challenges, criticisms, and difficulties with framing and defining PHC aside, interest in this form of care continues since its formalization in the Declaration. Although by the early 1990s emphasis had moved away from PHC in favour of other models of health care and achieving health for all (Haines et al. 2007), there has been a renewed interest in PHC since the early 2000s (Litsios 2004) including by the WHO (Haines et al. 2007). During the 1990s the health sector reform approach was popularized as a model for health care delivery in developing nations and was championed by the World Bank (Hall and Taylor 2003). Interestingly, the current renewed interest in PHC within developed nations in particular – including Canada, the UK, and New Zealand (Haggerty et al. 2007) – centres on reform and restructuring. In some instances PHC is being looked to as a model for providing restructured care. This is perhaps most evident in Canada where PHC is viewed as the foundation of the health care system and over one billion dollars have been spent since 2000 on reforming health care, and primary care in particular, by transitioning to a 'renewed' PHC system (Health Council of Canada 2005, 2008). Reforms specific to renewing PHC at the national and provincial/territorial scales have been explicated and are consistently cited as: (1) creating after-hours access to care; (2) implementing team-based care; (3) using new technology to enhance care (e.g., electronic medical records); and (4) changing physician payment models (Glazier 2008). This Canadian example illustrates one specific way in which PHC is gaining the attention of governments and health care systems now exactly 30 years after the conference in Alma-Ata was held.

Beyond a concept or idea, PHC is now not only a structural feature of health care systems worldwide, it is also generally understood and desired by the public the world over. Both of these factors mean that it is a fairly permanent feature of health and social care systems. In effect, to remove PHC and its approach to and philosophy of care from these systems would require policy and attitudinal shifts so fundamental that they are almost unimaginable. This can be said for few other post-World War II concepts in health care. Because of this importance, PHC has not only drawn the attention of policy-makers and health care administrators but also of researchers from a range of disciplines. Most obviously, health professional disciplines such as family medicine (e.g., Greenhalgh 2007) and nursing (e.g., Ross and Mackenzie 1996), to name just two, have engaged in research on PHC from its inception and have chronicled both its development and practice. Health service researchers from a number of disciplines have also been active in researching PHC. This work is featured prominently in journals such as the *Scandinavian Journal of Primary Health Care* and *Primary Health Care Research & Development* in addition to broader and discipline-specific health, health care, and social care publications. An explicit or implicit interest in PHC also exists within a number of other disciplines including those of the social sciences. This ranges from sociologists' long-standing interest in specific health professional groups (e.g., Armstrong et al. 1993) to other scholars' examinations of the political will to adopt comprehensive PHC (e.g., Werner and Sanders 1997). Further, as we show in this next section, geographers too have an interest in understanding and examining PHC.

**Introducing the Geographies of Primary Health Care**

Given what we have just discussed above, and how broadly PHC can be understood and interpreted, it is not surprising that geographers have had a long-standing interest in researching this form of care. However, with few exceptions, this work is more implicitly about PHC than explicitly. One of these exceptions is a collection of papers published in 2001 by the journal *Health and Social Care in the Community* edited by Tony Gatrell (2001) entitled 'Geographies of Primary Health-Care.' The collection showcases six articles focused on PHC in the UK context that most broadly get at issues of the organization and delivery of primary care and/or of uptake of and access to services. Gatrell acknowledges that the collection does not reflect the full range of PHC research within and beyond geography but contends that it does centre on a concern with 'place' as it relates to PHC at scales from the national through to the local. While the collection is primarily focused on what we have discussed above as primary care,[2] it does provide an initial formalization of the 'geographies of primary PHC' as an area of inquiry within the sub-discipline of health geography.

As noted above, there is little geographic inquiry explicitly situated within investigating PHC. For example, a search through the archives of *Health & Place*, the flagship journal of health geography, reveals only a very limited number of articles published within the last 13 years that reference PHC in their titles or abstracts.[3] There are, however, numerous articles that are relevant to PHC through a focus on primary care systems, on specific professional groups implicated in the provision of such care such as GPs, or even through their consideration of issues central to the principles of PHC. Because of this we believe the best way to explicate what we already know about the geographies of PHC is to introduce some of the concepts central to PHC that health geographers in particular have been most actively involved in investigating. In reviewing the principles of the Alma-Ata Declaration, we suggest that the concepts or topics of community/participation, access, and equity are those that health geographers have had long-standing interest in with regard to health and social care and their systems and services.

Ensuring equitable access to care is something that is enshrined in the Declaration signed at the conference in Alma-Ata. Haggerty et al. (2007, 304) contend that equity in PCH can be thought of as the 'extent to which access to health care and quality services are provided on the basis of health needs, without systematic differences on the basis of individual or social characteristics.' This

---

2   The contribution by Bailey and Pain (2001) serves as an exception. In this article the researchers examine the socio-cultural context of breastfeeding decision-making by women with a focus on health promotion strategies. They conclude by considering implications for policies that extend beyond health care. Such a study, while relevant to primary medical care, is clearly (though not explicitly in its framing) engaging in the type of PHC envisioned by the WHO at Alma-Ata.

3   In searching the *Health & Place* archives in late March, 2008 we found only ten articles that fit this criterion (i.e., referencing PHC in either the title or abstract). This is not intended to be thought of as a systematic or thorough review of the health geography literature. We instead use this point to assist in demonstrating the lack of explicit framing of health geography research as PHC-related.

definition is focused specifically on health care and so can usefully be extended to include social care more broadly along with certain development strategies in order to be more in keeping with a fuller understanding of PHC. Doing so allows for a broadened understanding of equity more along the lines of 'the fair and just distribution of resources' (Bowen 2001, 27). Attentiveness to issues of equity in relation to care involves investigating whether individuals experience difficulty in obtaining needed care, receive a lower standard of care, experience differences in interactions with care professionals, receive care that does not recognize their needs, or are generally less satisfied with the care received (Bowen 2001) due to social location, intersecting axes of difference, or other factors. Health geographers have a clear interest in equity as it relates to many issues including access to and receipt of the kinds of health and social care central to a PHC approach, particularly across space or between groups within a particular place (see, for example, Curtis 2004, and Chapter 2).

Equity as it relates to health outcomes is also central to the PHC approach. This relationship is demonstrated clearly in Broemeling et al.'s (2006) results-based logic model for PHC. In this model they identify various factors that both shape and are shaped by this form of care. Some of these factors are (Broemeling et al. 2006, 10):

- PHC inputs
  - o   Fiscal resources
  - o   Material resources
  - o   Health human resources
  - o   Policy
  - o   Governance structures and decision-making
- PHC ultimate outcomes
  - o   Sustainable health care system
  - o   Improved/maintained functioning, resilience, and health
  - o   Improved level and distribution of population health and wellness

The last factor under 'PHC ultimate outcomes' relates to a more equitable distribution of health and health outcomes. This listing serves as an important reminder that research focused on equity as it relates to PHC, including that by health geographers, need not focus exclusively on the distribution of health care resources across space and within place but also can consider the distribution of health, wellness, or even illness as it relates to the kinds of care provided using a PHC approach or within primary care systems.

Our above discussion of equity references notions of access. Access is certainly a topic of great interest to health geographers and other PHC researchers (see, for example: Feldman 2006; Field and Briggs 2001; Parker and Campbell 1998; Perry and Gesler 2000; Tsoka 2004; Wellstood et al. 2006). Wong et al. (2008) recently conducted focus groups with Canadians regarding PHC priorities, and spatio-temporal accessibility (i.e., timeliness and proximity) was among the six dimensions of care that were identified to be of most concern; this is but one example of the fact that access is also a topic of both interest and concern to health care consumers. As is demonstrated throughout the various chapters of this book, there are many ways of thinking about access in relation to PHC (Bowen 2001). While considering 'geographic access' to care (i.e., one's proximity to a location such as a clinic) is clearly overtly geographic, health geographers' interest in access extends far beyond this. Access can be further conceived of in relation to culture, economic status and resources, language, and with regard to PHC specifically we can understand that certain care professionals serve as gatekeepers to people's access to secondary and tertiary care. Barriers experienced by individuals can (in part or whole) make care inaccessible, which can include: (1) availability of services; (2) one's ability to pay for care; (3) needs-based barriers (e.g., language, awareness, physical accessibility); and (4) inequitable treatment (Bowen 2001). Issues of access as they relate to PHC extend well beyond an individual's ability to get to and/or use a particular form of care; access is also an important consideration with regard to communities' abilities to be involved in decision-making (i.e., access to power), nations' and systems' abilities to get the money and material goods needed in order to initiate and continue care delivery (i.e., access to resources), and even services' abilities to attract and retain qualified personnel in order to meet health human resources needs (i.e., access to workers).

In looking at the non-binding Declaration signed at the Alma-Ata conference, the centrality of 'community' in relation to PHC is clear, in addition to the way it is linked to participation. Haggerty et al. (2007) identify four community-oriented dimensions of PHC provision through primary care systems: (1) client/community participation; (2) equity; (3) intersectoral team practice; and (4) population orientation. This first dimension in particular is illustrative of the connection between community and participation in PHC. The second dimension references our points made above, which is that a community's access to certain forms of power exemplifies a more equitable distribution of control over decision-making. Interest in issues of community/participation more broadly can bee seen in the work of human geographers, including health geographers specifically. The roles of communities in hosting sites of care, in care delivery, in building capacity, in bringing about change, in responding to need(s), and in creating their own local cultures of care are all relevant to health geographers' inquiry regarding PHC and primary care systems. The participation of community members in each of these, whether as formal paid workers or individual citizens, is also of relevance to geographers and non-geographers alike. Examples of such work include Charlesworth's (2001) investigation of the development of partnerships within UK primary care provision including between government sectors and community-based voluntary providers, Janes' (2004) critique of community participation in PHC using the case of health care reform in Mongolia whereby she places emphasis on how local and global are connected through health care policy, and Prentice's (2006) quantitative analysis of the impacts of neighbourhood on primary care provision using the case of Los Angeles, California where it is argued that local health norms and care availability are significant variables within communities that shape individual uptake of primary care.

## Book Overview

Our purpose in bringing this edited collection together is to compile significant contributions that inform how we can think geographically about health care and specifically PHC. As noted above, a goal is to showcase this work through interrogating the landscapes that inform the very principles of PHC and ultimately shape how this care is both delivered and received (we revisit this in Chapter 15). In the chapter that follows we provide an overview of key developments in the geographies of health care and the current state of research on this topic within health geography. The main part of the book is divided into three main sections, discussed below. Importantly, considerable overlap exists between these sections in terms of what the chapters present, thereby demonstrating how these themes (i.e., practice and delivery, people, and places and settings) interrelate. Following this, a final chapter draws together important themes of the collection and puts forward an agenda for continued research in the geographies of PHC. It should be noted that no single framing or definition of PHC is used throughout the chapters and so there is (useful) diversity in how the authors approach and understand such care.

### Practice and Delivery

In this section of the book the authors focus explicitly on core geographic concepts as they relate to the practice and delivery of PHC. By practice we are not referring here to clinical practice but rather the practices through which care gets developed and delivered. Neil Hanlon considers the concepts of access and utilization as they relate to PHC in Chapter 3. He argues that we need to consider PHC more broadly, rather than primary care specifically, in order to more accurately understand access and utilization as doing so allows us to better account for equity in service distribution and use. The chapter concludes by calling for a 'multidimensional perspective' on access and utilization by health geographers and others. Nadine Schuurman's chapter follows (Chapter 4) and serves as an interesting response in that she focuses on specific dimensions related to PHC access. In it she provides a case example regarding access to specific PHC services (i.e., rural maternity and trauma care) in the Canadian province of British Columbia. She contends that dimensions such as population density, physical distance, and socio-economic status must be accounted for when considering access to PHC. In doing so she uses GIS techniques to identify ways to overcome distance and lessen rural vulnerability in the case example.

Janine Wiles and Mark Rosenberg contribute Chapter 5. In it they examine the geographic concept of scale as it relates to PHC. They start by providing a very useful overview of scale in human geography. They then move to consider the relationship between PHC and scale. To further evidence this relationship they provide examples of scalar concepts embedded in the Alma-Ata Declaration and also Canadian PHC. They conclude by highlighting the implications of scale and how it is conceived and enacted for PHC provision. The chapter that follows (Chapter 6) is authored by Leah Gold and provides a local-scale investigation of the roll-out of a new PHC initiative in Perú. Using an ethnographic approach that includes interviews with health care providers/administrators, she questions whether the initiative is more reflective of selective PHC although it is being touted as a comprehensive approach. She places significant scrutiny on how community involvement is conceived of and enacted in this initiative, arguing that rural peasants are mostly excluded from participating in decision-making and often times from even receiving care.

*People*

Issues of geography, space, and place are central to how PHC is lived out and experienced both by those who give and receive such care. In this section of the book the focus is on specific provider groups, though through discussing these groups the implications for PHC consumers and others (e.g., community members) are clear. Two chapters focus mostly on GPs and the care they deliver through primary care systems somewhat informed by a PHC approach. Gina Agarwal examines geographic aspects of GPs' practice in Chapter 7. Her discussion is organized around the *Four Principles of Family Medicine* and makes note of how both community and environment are important to practice. Interestingly, in the conclusion she moves to reflect on the spatiality of her own practice as a GP as it relates to the discussion she presents. Ross Barnett and Pauline Barnett also consider GPs and their practice in Chapter 9. Specifically, they examine the reinvention of primary care in New Zealand, including changes to how care is organized and the implications for GPs. They place their discussion of New Zealand in an international context by comparing elements of primary care reform or reinvention across three nations.

   The other two contributions to this section of the book focus on different provider groups and are united in their perspective that biomedical dominance within the health care system must be overcome in order to more fully effectuate PHC. Jennifer Lapum and colleagues focus on nursing practice in Chapter 8. In doing so, they critically reflect on the 'place' of nursing in PHC. Considering PHC as both a philosophy and practice, they contend that nurses' roles in such care have gone unnoticed and under-recognized for many reasons. They conclude by calling for an increased recognition of nurses' roles in PHC and also for nurses to play a more central role in PHC decision-making, including at the international scale. Daniel Hollenberg and Ivy Bourgeault focus on the practice of complementary/ alternative and traditional medicine (CAM/TM) in Chapter 10. They start by contextualizing PHC through a discussion of the Alma-Ata Declaration and explain the role for CAM/TM in PHC. Through a case example of a specific integrative health care setting (i.e., where CAM/TM and biomedicine are provided in the same clinic), they argue that integrative health care provides new spaces for the provision of CAM/TM and ultimately PHC.

*Places and Settings*

Examining the space of care provision and larger settings that inform the practice and delivery of PHC are topics central to the development of a geographic perspective on PHC. In the final major section of the book the focus is on notions of place and the settings in which care is delivered. The first two chapters turn to the community clinic as a care site. Valorie Crooks and Gina Agarwal consider the role of the clinic environment in care provision and present an original case example based on interviews conducted with women experiencing clinical depression in order to do so in Chapter 11. They draw on issues such as the purposes of the multiple micro-spaces within the clinic, internal design and layout, and people's different 'readings' of the clinic space to support their contention that the environment matters with regard to how care is both delivered and received. They conclude by considering the implications of their discussion for implementing the kinds of PHC envisioned in the Alma-Ata Declaration. Robin Kearns and Pat Neuwelt contextualize their consideration of the clinic with a more nuanced understanding of the role this place plays within the community in Chapter 12. Drawing on their work in New Zealand, they focus on the role of community participation in the clinic and how the very nature of this place can assist with distinguishing between primary care participant and consumer. They further consider how to best actualize community participation in the New Zealand context and conclude by calling for health geographers to place increased focus on how local clinics can and do enact community participation.

The last two contributed chapters closely examine the roles of and need for PHC in different care settings. In Chapter 13 Nicole Yantzi and Mark Skinner turn to the home as a site of PHC giving and receiving. They draw on their own existing research to illustrate the many roles that providers in the home play, including that they often initiate first-contact care and enhance continuity of care. Their contention is that the roles of these providers are undervalued but yet are essential to the continuum of PHC provision. In concluding their chapter they call for greater attention to be paid to those who provide care in the home and their needs. David Conradson and Graham Moon, in Chapter 14, also focus on an under-recognized group related to PHC provision: consumers living 'on the street' (e.g., homeless individuals, rough sleepers). They frame this group as being difficult to reach but also in need of care and use two case examples to illustrate whether and how NHS-funded Walk-in Clinics and voluntary organizations are able to meet their needs. Their chapter gets at issues of equity of access in PHC provision and they conclude by suggesting that mobile care provision may be the best way forward in terms of meeting the needs of this group.

**Acknowledgements**

We are thankful to all those who have contributed chapters to this book. Without their efforts we could never have undertaken this project. We are also appreciative of Ashgate and the series editors for their support in seeing this project through to completion. Finally, we must acknowledge the assistance of Melissa Giesbrecht who worked with us on reviewing several of the chapters for formatting.

**References**

Armstrong, P., Choiniere, J. and Day, E. (1993), *Vital Signs: Nursing in Transition* (Toronto: Garamond Press).

Bailey, C. and Pain, R. (2001), 'Geographies of Infant Feeding and Access to Primary Health-care', *Health and Social Care in the Community* 9:5, 309–17.

Bowen, S. (2001), 'Access to Health Services for Underserviced Populations in Canada', in *"Certain Circumstances" Issues in Equity and Responsiveness in Access to Health Care in Canada* (Ottawa, ON: Health Canada).

Broemeling, A.-M., Watson, D.E., Black, C. and Reid, R.J. (2006), *Measuring the Performance of Primary Health Care – Existing Capacity and Future Information Needs* (Vancouver, BC: Centre for Health Services and Policy Research).

Canadian Nurses Association (CNA) (2005), *Primary Health Care: A Summary of the Issues* (Ottawa, ON: CNA).

Cardarelli, R. and Chiapa, A.L. (2007), 'Educating Primary Care Clinicians About Health Disparities', *Osteopathic Medicine and Primary Care* 1:5.

Charlesworth, J. (2001), 'Negotiating and Managing Partnership in Primary Care', *Health and Social Care in the Community* 9:5, 279–85.

Cueto, M. (2004), 'The Origins of Primary Health Care and Selective Primary Health Care', *American Journal of Public Health* 94, 1864–74.

Cueto, M. (2005), 'The Promise of Primary Health Care', *Bulletin of the World Health Organization* 83:5, 322–3.

Curtis, S. (2004), *Health and Inequality* (London: Sage).

Feldman, R. (2006), 'Primary Health Care for Refugees and Asylum Seekers: A Review of the Literature and a Framework for Services', *Public Health* 120:9, 809–16.

Field, K.S. and Briggs, D.J. (2001), 'Socio-economic and Locational Determinants of Accessibility and Utilization of Primary Health-care', *Health and Social Care in the Community* 9:5, 294–308.

Gatrell, A.C. (2001), 'Guest Editorial: Geographies of Primary Health-care: Perspective and Introduction', *Health and Social Care in the Community* 9:5, 263–5.

Gesler, W.M. and Kearns, R.A. (2002), *Culture/Place/Health* (London and New York: Routledge).

Glazier, R. (2008), 'Mapping the Future of Primary Healthcare Research in Canada', *Annals of Family Medicine* 6:1, 89–90.

Green, A. (1987), 'Is There Primary Health Care in the UK?', *Health Policy and Planning* 2:2, 129–37.

Greenhalgh, T. (2007), *Primary Health Care: Theory and Practice* (UK: BMJ Books).

Haggerty, J., Burge, F. and Levesque, J.-F. et al. (2007), 'Operational Definitions of Attributes of Primary Health Care: Consensus Among Canadian Experts', *Annals of Family Medicine* 5, 336–44.

Haines, A., Horton, R. and Bhutta, Z. (2007), 'Primary Health Care Comes of Age. Looking Forward to the 30th Anniversary of Alma-Ata: Call for Papers', *The Lancet* 370, 911–3.

Hall, J.J. and Taylor, R. (2003), 'Health For All Beyond 2000: The Demise of the Alma-Ata Declaration and Primary Health Care in Developing Countries', *Medical Journal of Australia* 178, 17–20.

Health Council of Canada (2005), *Primary Health Care – A Background Paper to Accompany Health Care Renewal in Canada: Accelerating Change* (Toronto, ON: Health Council of Canada).

Health Council of Canada (2008), *Fixing the Foundation: An Update on Primary Health Care and Home Care Renewal in Canada* (Ottawa, ON: Health Council of Canada).

Hutchison, B., Ableson, J. and Lavis, J. (2001) 'Primary Care in Canada: So Much Innovation, So Little Change', *Health Affairs* 20:3, 116–31.

Janes, C.R. (2004), 'Going Global in Century XXI: Medical Anthropology and the New Primary Health Care', *Human Organization* 63:4, 457–71.

Litsios, S. (2002), 'The Long and Difficult Road to Alma-Ata: A Personal Reflection', *International Journal of Health Services* 32:4, 709–32.

Litsios, S. (2004), 'Primary Health Care, WHO and the NGO Community', *Development* 47:2, 57–63.

Mohan, J. (2000), 'Geographies of Health and Health Care', in R.J. Johnston, D. Gregory, G. Pratt and M. Watts (eds), *The Dictionary of Human Geography*, 4th edition (Oxford and Massachusetts: Blackwell Publishers Ltd.).

Pappas, G. and Moss, N. (2001), 'Health For All in the Twenty-first Century, World Health Organization Renewal, and Equity in Health: A Commentary', *International Journal of Health Services* 31:3, 647–58.

Parker, E.B. and Campbell, J.L. (1998), 'Measuring Access to Primary Medical Care: Some Examples of the Use of Geographical Information Systems', *Health & Place* 4:2, 183–93.

Perry, B. and Gesler, W. (2000), 'Physical Access to Primary Health Care in Andean Bolivia', *Social Science & Medicine* 50:9, 1177–88.

Prentice, J.C. (2006), 'Neighbourhood Effects on Primary Care in Los Angeles', *Social Science & Medicine* 62:5, 1291–1303.

Ross, F. and Mackenzie, A. (1996), *Nursing in Primary Health Care: Policy and Practice* (London and New York: Routledge).

Schoen, C., Osborn, R. and Huynh, P.T. et al. (2004), 'Primary Care and Health System Performance: Adults' Experiences in Five Countries', *Health Affairs*, Supplement Web Exclusives W4, 487–503.

Taylor, C. and Jolly, R. (1988), 'The Straw Men of Primary Health Care', *Social Science & Medicine* 26:9, 971–7.

Tejada de Rivero, D.A. (2003), 'Alma-Ata Revisited', *Perspectives in Health Magazine* 8:2.

Tsoka, J.M. (2004), 'Using GIS to Measure Geographical Accessibility to Primary Health Care in Rural South Africa', *South African Journal of Science* 100:7–8, 329–30.

Wakai, S. (1995), 'Primary Health Care Projects and Social Development', *The Lancet* 345, 1241.

Walsh, J.A. (1988), 'Selectivity Within Primary Health Care', *Social Science & Medicine* 26:9, 899–902.

Watson, D.E., Krueger, H., Mooney, D. and Black, C. (2005), *Planning for Renewal – Mapping Primary Health Care in British Columbia* (Vancouver, BC: Centre for Health Services and Policy Research).

Wellstood, K., Wilson, K. and Eyles, J. (2006), '"Reasonable Access" to Primary Care: Assessing the Role of Individual and System Characteristics', *Health & Place* 12, 121–30.

Werner, D. (1995), *The Life and Death of Primary Health Care*, presentation given for the International People's Health Council at the NGO Forum, United Nations 'Global Summit', March 7, 1995.

WHO (World Health Organization) (1978), *The Alma-Ata Declaration* (Geneva: World Health Organization).

WHO-EM (World Health Organization Regional Office for the Eastern Mediterranean) (2003), *Technical Paper, Primary Health Care: 25 Years After Alma-Ata*. Fiftieth Session, Regional Committee for the Eastern Mediterranean.

Wisner, B. (1988), 'GOBI versus PHC? Some Dangers of Selective Primary Health Care', *Social Science & Medicine* 26:9, 963–9.

Wong, S.T., Watson, D.E., Young, E. and Regan, S. (2008), 'What do People Think is Important About Primary Healthcare?', *Healthcare Policy* 3:3, 89–104.

**Internet-based References**

Health Canada (2006), 'What is Primary Health Care?' Available online at http://www.hc-sc.gc.ca/hcs-sss/prim/about-apropos_e.html. Accessed March 5, 2008.

WHO-SEA (World Health Organization Regional Office for South-East Asia) (2006), 'The Fourth Decade'. Available online at www.searo.who.int. Accessed February 16, 2008.

Chapter 2

# Geographical Perspectives on Health Care: Ideas, Disciplines, Progress

Gavin J. Andrews and Valorie A. Crooks

This chapter reviews geographical research on health care. Its general message is that this is a wide-ranging academic field involving substantial disciplinary contributions from the health sciences and the social science of geography. Two differences are however highlighted in the nature of these contributions. The first relates to the health sciences being subject-orientated, whilst geography being perspective-orientated. In the health sciences, the engagement with health care has been a necessary constant, whilst concern for geography has been occasional and varied. In the discipline of geography, geographical inquiry has been a necessary constant, whilst concern for health care has been occasional and varied. The second difference relates to longevity and leadership. Geographical ideas on health care have by far the longer history in the health sciences which forged early inquiry from ancient Greek times and for many centuries thereafter. In contrast Geography has only led innovations from the late nineteenth century and during twentieth century, but has introduced a vast range of new ideas in this comparatively short time. More recently, the health sciences have returned to a leadership role by forging new geographical literatures and directions, particularly with regard to work issues.

The chapter is not focused on primary health care (PHC) alone. Instead, by engaging with health care in its broadest sense, the intention is to give readers an overview of geographical perspectives so that they might be able to locate the specific arguments presented in the following chapters in a broader tradition. We use the word geographical as an umbrella term to denote research on health care with a conceptual emphasis on space and/or place as broadly defined, and/ or research that engages directly with phenomena such as nature, landscapes, environments, regions and urban or rural areas.

## Geographical Perspectives on Health Care in the Health Sciences

In this section we provide an evolutionary overview of geographical inquiry into medicine and health care by non-geographers. For the most part these scholars have been health scientists of various types, and we consider how they have advanced our knowledge of the complex connection between place/space and health care.

*Early Medicines (400BC–1950s)*

The very earliest human civilizations believed that a god or gods controlled everything observable in the world. Diseases were often thought of as punishments on individuals and populations, handed down as moral judgments on their behaviours. Only when alternatives to exclusively supernatural causes of disease were accepted by societies did spatial/geographical explanations emerge (Barrett 2000). Early geographical perspectives were not employed by those outside the practice of medicine as they are predominantly today (such as social scientists and health service researchers) but instead were taken on by medicine itself and medical thinkers of the time (albeit they would not have been recognized as geographical).

The origins of a geographical perspective can be traced to the origins of both medicine and academic thought in ancient Greece. In 400BC Hippocrates' *Airs, Waters and Places* introduced a number of precise, authoritative and direct geographical ideas (Barrett 2000). Speaking to the traveling physician, Hippocrates argued that whomever wished to investigate medicine properly should consider a range of issues particular to his or her locality including seasons, climate, water quality, and inhabitants' pursuits (eating, drinking, exercise, work). (Barrett 2000). In later centuries, ideas from *Airs, Waters and Places* were used by other societies and were combined with their own ways of explaining the world. Roman Civilizations, for example, used ideas from *Airs, Waters and Places* alongside their own theories on hygiene and health (Barrett 2000).

At approximately the same time as *Airs, Waters and Places*, geographical concepts emerged in ancient Chinese and Indian medicine (acknowledging our section title 'health science' is perhaps an incorrect descriptor in these cases). Written in the second century BC, the key Chinese documents *Huang Ti Ching*, *Su Wen* and *Ling Shu Ching* introduced the idea of parallelism, which included the concepts of macrocosm (outer world) and microcosm (body inner world). As Barrett (2000) suggests, patterns in the outer world – such as features of landscape – were thought to be reflected in the inner world. This led to the explanation, for example, that river systems and veins were parallel drainage systems (Barrett 2000). These documents also included theories that connected five elements – metal, wood, water, fire, soil – to symptoms of illness in specific geographical locations. Many associated connections between care and place were made including, for example, that coastal fish diets cause ulcers best treated by acupuncture (Barrett 2000). In India, the dominant system of medicine was Ayurveda. Two works here, *Charaka Samhita* and *Susruta Samhita*, like *Airs, Waters and Places*, talked about location impacting on disease through rivers, moisture and temperature (Barrett 2000).

Although geographical ideas were firmly rooted in the above times and places, it should however be noted that many later societies either did not retain them or never adopted them. Early Christian societies, for example, during 'the age of faith' preferred metaphysical explanations for disease; their ideas about 'god's will' and disease as punishment being drawn predominantly from the Old Testament (Barrett 2000). In later centuries, European exploration and colonization led to a reemergence of spatial ideas in medicine. Maritime exploration, although motivated by economic, cultural and political factors, also included a quest for knowledge of 'exotic' health conditions. As Barrett (2000) describes, Cadamosto of Portugal voyaged to Africa in 1455 and recorded native consumption of salt, interpreting this as a way to prevent blood from putrefying in the heat. In the sixteenth century more expansive commentaries developed within two themes: (1) diseases in new lands that could affect explorers and colonists, and (2) traditional medicines used by 'natives' which could be returned home as remedies (Barrett 2000).

In the later years of the age of exploration Western medical ideas, service models and basic facilities were exported to colonized societies. Here medicine was involved centrally; particularly as from the early 1700s physicians increasingly sailed on voyages of discovery and developed new disciplines such as tropical medicine (Barrett 2000). As exploration developed, Europeans carried diseases of European origin with them, and also moved African disease elsewhere, both of which decimated populations, and ultimately and ironically gave scholars more to study (Barrett 2000). Underpinning much medical thought at the time was the theory of environmental determinism. This involved an understanding that local environmental conditions determine the character of people, their activities, behaviour and physical characteristics. Such arguments reinforced white European supremacy in the world order and were used to explain European 'achievements' over peoples living under different environmental conditions in other areas of the world. In terms of medical thinking, the theorist Bodin, for example, wrote in 1566 that people from the Mediterranean were, amongst other things, dry, hard and weak, whilst people from Baltic were warm, large and robust (Barrett 2000).

During the eighteenth century and thereafter greater meanings were attributed to space and place in medical thought, including more attention being paid to disease in regional monographs, to the aerial patterning of disease, and also on how to treat geographically-specific conditions (Barrett 2000). As Barrett (2000) explains, key works published during this century by physicians included Hoffman's (1705) *Diseases Peculiar to Certain Peoples and Regions*, which associated climate, disease and nutrition with medical care in places; Lind (1768) which explored health in tropical colonies and treatment solutions; Dehorne (1786) which explored the geography of disease in France through consideration of the health of soldier garrisons in certain regions; Halle (1787) on Africa and hygiene; and Finke (1792–1795) which presented global explanations of health and places. During the nineteenth century research was increasingly informed by Darwinian thought and involved a greater appreciation of evolutionary paths and natural histories of peoples and diseases in specific places. Key works of note here included Drake (1850) which focused on principal diseases in different ethnic groups in the North-West of the United States of America (Barrett 2000).

Also particularly influential at the time was Florence Nightingale's (1859) *Notes on Nursing: What it is and What it is Not*, which included many geographical ideas ranging from connecting urban living conditions and disease, to comments on clinical environments – and the need for correct heat, light, air, water and interactions and behaviors in patients' rooms. The latter observations were important, being the first time that human influences on, and features of, places were recognized as being critical to health care. Since the latter part of the twentieth century, building on Nightingale's work, health professional disciplines have developed a growing awareness of how environment matters at the micro-scale to ways of working and caring. The meta-paradigm of 'nursing environment', for example, has helped define and model the profession for over five decades (Andrews and Moon 2005; Fitzpatrick and Whall 1983; Kleffel 1991; Thorne et al. 1998).

Late, in the twentieth century, studies of the geography of health and disease developed substantially in the health sciences and included the establishment of specialist disciplines/approaches such as geographic pathology, geomedicine, medical geology and spatial epidemiology. The latter of these, for example, places geographical location as an important explanatory variable for the incidence and spread of disease.

*Health Services Research (1960–2008)*

Over the past three decades a spatial perspective has remained in the health sciences, though the context for understanding the place-health-care relationship has changed considerably. Since the late 1970s, health services research has developed into a vast multidisciplinary field that includes contributions from a range of health professional disciplines, biostatisticians, epidemiologists and others. Part of the development of this meta-field has also involved the infusion of a range of social science perspectives including from psychology, sociology, political science, economics and, most relevant to this discussion, geography. Each of the above has brought its own unique ideas, creating a mosaic of disciplines. Alongside an extension of disciplinary perspectives has been an extension of the subjects studied to include a wide range of empirical and conceptual themes such as:

- health care delivery models
- theories and concepts in health care
- financial and other resource issues
- management, administration and governance

- regulation, quality and safety issues
- market structures and relationships
- the development of specific categories of services
- health professions and human resources
- technological developments
- population health needs, outcomes and expectations
- facility planning and design

Across the breadth of health services research, geographical perspectives employed vary greatly, ranging from research engaging very specifically with how health and health care are patterned at national and sub-national levels to studies which collapse spatial concerns into broad categories such as rural services, urban services, or classifications such as 'the Northeast', 'remote area' or 'inner city'. Otherwise geography might arise through a focus on a specific country, region (e.g., Baron-Epel et al. 2005, 2007) or internal political jurisdiction such as Canadian provinces or British counties (e.g., Fell et al. 2007). With respect to region, the broader regional structural/administrative contexts to health care – including regionally devolved power and regional models of care delivery – are an increasingly important consideration in mainstream health services research (see Secker et al. 2006). Meanwhile, at the micro-scale, common classifications used in inquiry such as hospitals, clinics and wards equally imply some sort of geographical form. In all these ways a spatial perspective might be evident but remain a low priority in analysis, typically referring very basically to the location of the service or a research site.

Building on this implicit spatial component of health services research has been a more direct and explicit engagement with geographical concepts and theories during the past ten years. Alongside policy makers and civil servants, health services researchers have recognized place and space as important to service provision and population health. Such change is reflected, for example, in a recent health policy research bulletin published by Health Canada (2007, 1) which states: 'Research from several disciplines is currently shedding new light on the complexities of the relationships between a place, its people and their health.' In approaching place, the bulletin states (Health Canada 2007, 7):

> ... depending on the field of study, one can consider place from various perspectives – climatic zones, regions with similar topography, areas under the same political control, to name a few. As a starting point for studying people, place and health, place can be thought of as a geographic area where men, women, boys and girls live in all their diversity.

This particular publication goes on to distinguish between local economic environments (e.g., paid employment, economic status, business ownership), physical environments (e.g., natural, built), public policy environments (e.g., citizen participation, legislation, zoning laws) and social environments (e.g., opportunity, leisure, participation, neighbourhood and community, crime and other problems, issues specific to specific social, demographic and ethnic groups). It also showcased recent inquiry in health services and policy research which defines, measures and analyzes health and place (Gupta and Senzilet 2007), and specifically considers topics including population patterns and health (Creasey 2007), health variations and disparities (Gupta and Ross 2007), rural health and health care issues (DesMeules et al. 2007), the urban built environment (Grenon et al. 2007) and pollution (Cakmak 2007). Beyond this specific publication, the growing explicit spatial awareness in health services research surfaces in the pages of the many journals, books, websites and other venues that publish work in this wide-ranging field.

*Health Professional Research (1990–2008)*

Following the initial work of Nightingale noted above, during recent decades concepts such as 'clinical environment', 'therapeutic environment' and 'quality of environment' have become centrally located in debates on improving practice in many professional and clinical specialties (Andrews and Evans 2008). Often, for example, these concepts are related to focused discussions on facility design and quality of life (Brawley 2001).

Spatial perspectives in health professional research have however developed simultaneously in another way. As documented in a number of recent reviews, nursing research in particular has undergone somewhat of a 'geographical turn' during the last decade, meaning that scholars have drawn explicitly on ideas, literatures and concepts in human and health geography to explain aspects of nursing (Andrews 2006; Andrews and Evans 2008; Carrolan et al. 2006). As Andrews and Shaw (2008) tell us, five broad categories of geographical studies of nursing can be identified. First, there are those studies that without engaging directly with clinical practice itself, map important geographical contexts to practice. They might, for example, consider the spatial features of population and patient groups, diseases or features of communities (Skelly et al. 2002) or very practically explore the potential of geographical tools such as computer cartography and GIS (Moss and Schell 2004).

A second group of studies articulates the dynamics between spaces and places and the character of specific professional specialties of nursing (Andrews and Shaw 2008). Cheek (2004) and Andrews et al. (2005), for example, consider various places for older people's care and Bender et al. (2007) conceptualize community in community health nursing. A crosscutting argument exists in this research that space and place are central to the advancement of these specialties and the care they provide (Andrews and Shaw 2008).

Third, attention has been paid to spatial career movements and the cultural, social and economic forces that shape them at local (Brodie et al. 2005), national (Radcliffe 1999), cross-national (Pratt 1999) and global (Kingma 2006) scales (Andrews and Shaw 2008). Radcliffe (1999), for example, considers disparities in career advancement and their relationship to relocation flexibility. Kingma (2006), meanwhile, considers individual and group features of workforce migration in the global health care economy. This research ties into the considerable literature on workforce and human resources that it a central pillar in nursing research.

A fourth group of studies articulates how places reflect and impact upon professional-patient and inter-professional interactions and relationships (Andrews and Shaw 2008). Whilst certain scholars have based their analyses on traditional institutional settings, for example highlighting spatial and moral proximities in wards (Malone 2003), others have deconstructed the meaning of home in homecare (Gilmour 2006) and the moral issues specific to that place (Liaschenko 1994, 1996, 1997, 2003; Peter 2002; Peter and Liaschenko 2004). Meanwhile consideration has been given to the roles of technology in spatial relationships (Purkis 1996), and the gendered 'playing out' of physical interactions, inter- and intra-professional rivalries and networking in institutions (Halford and Leonard 2003, 2006a, 2006b; West and Barron 2005).

A fifth group considers the roles of space and place in the production and uptake of clinical evidence (Andrews and Shaw 2008). Whilst certain studies directly compare practice and care settings as the controlled variable in clinical trials (Hodnett et al. 2005; McKeever et al. 2002), critical studies discuss how clinical trials – and the potential generalizability of their findings – are impacted by unique place-specific personal, professional, technical and interpersonal influences (Angus et al. 2003). Building on this are reflections on the translation of research knowledge from journals into practice settings, and the existence of situated obstacles and facilitators (Andrews et al. 2005).

The collective contribution of these five categories of recent nursing studies has been to add a geographical perspective on work roles, processes, experiences and concepts. As we shall see in the following section, health geography has not adequately covered this ground, and nursing studies are currently filing a distinct gap in the literature (see also Andrews and Evans 2008).

## Studying Health Care in Geography

In this section we provide an overview of geographers' research on health care. By comparison to the geographical perspectives in medicine and the health sciences, discussed above, this has been a relatively recent undertaking.

*Traditional Perspectives (1960–2008)*

As Barrett (2000) outlines, early commentaries by geographers were published sporadically from the early nineteenth century onwards. Humboldt (1807), for example, focused on the location of Yellow Fever, Ritter (1832) on the effects of smoking opium, and Johnston (1856) on distributions of health and disease. Only in the 1940s, however, did geographers start to attend to disease and health, substantively developing 'medical geography' (now termed health geography) as a recognizable sub-discipline of human geography.

Mirroring the early studies noted above, a central and longstanding field of inquiry in the sub-discipline explores diseases. Much of this work uses a positivistic approach in explaining their distributive features as they relate to social phenomena (Gatrell 2002). Although the connections made in this research to health care are often indirect, as Andrews and Evans (2008) point out, knowledge of when and where diseases occur and how they spread is of great relevance to health care in terms of developing good prevention and care/cure services. Many examples exists of how this might be so. Gatrell (2002), for example, notes work conducted in the 1990s that demonstrated how AIDS is spread spatially by both *contagious diffusion* (i.e., carriers to others through their close spatial proximity) and through *hierarchical diffusion* (i.e., flows of people in transport routes 'leapfrogging' the disease from place to place) (see Gould and Wallace 1994). In this case, such geographical information is critical to planning effective public health and medical responses to AIDS.

Meanwhile another field of the sub-discipline has exhibited a far more direct engagement with health care. Here commentators have described the aerial patterning of health care and its relationships to service use and patterns of health and disease, ranging from international and national to localized scales (Eyles 1990; Mohan 1998). As Meade and Earickson (2000) explain, concerns for inequity in service provision in the sub-discipline have motivated much of this research, reflected by sustained attention to what geographers refer to as 'underserviced' populations and places. In working towards solutions, geographers have often attempted to calculate the most equitable spatial allocations of health care. Over time, and particularly since the 1970s, accessibility and utilization of health care have become major concerns in the sub-discipline (Joseph and Phillips 1984). As Meade and Earickson (2000) describe, here access is considered a product of four variables:

- The availability of services, including their existence and distance. (Commentators have highlighted distance-decay in use, sometimes referred to as 'the friction of distance')
- Means of access (including cost and income, insurance)
- Non-discriminatory systems, services and professionals
- Attitudes and knowledge of consumers (including different health seeking behavior and health preferences).

Ultimately, as Jones and Moon (1987) observe, for geographers 'territorial justice' is reached when in all places need is met equally by an appropriately proportioned amount of resources. Although, as Meade and Earickson (2000) note, on a practical level territorial justice is difficult to achieve under any circumstances – the most obviously problematic being the global scale. Consequently, as part of a more reflective approach, the sub-discipline has moved away from automatically supporting territorial justice. For example, not all scholars consider the unequal spatial distribution of health care to be a problem, particularly those who believe in the sanctity of the marketplace and its ability to determine where resources are located (Meade and Earickson 2000). Such a perspective also allows for the prioritizing of equity in health service access over equality which is a topic that has garnered significant attention by health geographers.

*Traditional Theories and Methods of Inquiry*

Underpinning much inquiry into health care in health geography has been a number of well-used theories, concepts, methods and analytical approaches. Three of these are described and critiqued briefly below to provide a flavor of research and how it has progressed. The first example is central place theory (CPT), which has been used when calculating the optimal spatial organization of health care delivery. CPT assumes that ideally there should be a hierarchy of services which resembles a pyramid. On the top, a small number of very large facilities would have a broad spatial coverage, a wide scope of services, specialist services and higher order services. Underneath these, consecutively smaller facilities would be more numerous, have smaller spatial coverage, narrower ranges of services, lower order services and increasingly generalist services (Joseph and Phillips 1984). The limitations of such stylized theories and models are, however, recognized. As Meade and Earickson (2000) suggest, CPT does not account for 'irrational' and inconsistent provider and consumer behavior, physical barriers that change through space (such as transportation), social difference and inequality or the inability of such a system to cover very remote populations. All these limitations mean that any system mirroring the model might not be used in the way it was intended (to provide maximum coverage). Hence, the reality is that, despite academic attention to CPT, health care systems rarely resemble the stylized model, which has more potential as a prescriptive tool (Meade and Earickson 2000).

The second example is Location-Allocation Modeling (LAM), which is used to advise the planning of services and has an equally long pedigree in health geography (Gould and Leinbach 1966; Morill and Earickson 1969). Like CPT, it is an exercise undertaken with the intent to redress spatial imbalances in resource allocation. As Gatrell (2002) notes, these models both propose both facility location criteria and allocations of people to facilities that typically minimize distance (Bailey and Gatrell 1995; Smallman-Raynor et al. 1998). The critique of LAM was led by Rosenberg (1988) who questioned its implicit assumptions about consumer and supplier behavior. These include, for example, that individuals will travel to the nearest health care setting and longer to larger settings, that demand centers around one single point and that services at each point are similar (Meade and Earickson 2000; Rosenberg 1988). Hence, Rosenberg sees LAM as theoretically optimal in a perfect regular world that unfortunately never exists. Moving the debate forward, Rosenberg suggested an alternative to LAM that places individuals, as opposed to services, at the centre of decision-making. Here individuals are influenced by:

socio-cultural factors (e.g., community, peers, family) and political-economic factors (e.g., government policy, health and informal care) (Meade and Earickson 2000; Rosenberg 1988).

The third example of a traditional mode of inquiry is the use of Geographical Information Science (GIS) which has facilitated CPT and LAM analyses – and other types – and, most importantly has allowed ever greater degrees of analytical sophistication and broader application. GIS originates from computer mapping and cartography in human geography in 1960s, but is now used across many of its sub-disciplines including health geography. It includes a range of computer-based mapping tools (named Geographical Information Systems when referring to specific applications) which help manage, store, retrieve, integrate, analyse and display data relating to health or health care when locations can be assigned (Meade and Earickson 2000). These systems range from simple pc-based software packages to dedicated expensive and complex hardware. Because GIS digitizes data, it can be manipulated into easily accessible and aesthetic forms. Broad areas of application include mapping environmental exposure data, rates of illness, facilities and the use of facilities and health seeking or avoidance behaviour (Meade and Earickson 2000). GIS can be used from the scale of countries to the scale of neighborhoods whilst typical categories of plotted information include densities and clusters, rates, and single point distributions of health and health care phenomena. As Gatrell (2002) describes, when applied to health care, one example might be mapping locations of facilities (and their various attributes such as staff and services) along with the residential locations of patients treated there (and their various attributes such as demographic qualities and health status). This might help researchers draw conclusions on the dynamics between production and consumption.

*Recent Developments in Health Geography (1993–2008)*

Recent changes and developments in the sub-discipline have shaped and informed geographers' inquiries on health care. In an influential paper, Kearns (1993) called for radical sub-disciplinary reform, including a much richer conceptualization of place. Kearns suggested that (medical) geographers had previously paid little attention to place in health care delivery, typically referring place only as the location of services and research sites. Kearns noted how places are far more complexly constructed social phenomena that hold particular significance for people, how a person's background and experience may shape their experience of places and how places affect people's opportunities and activities. Indeed, experiences of places are thought to be obtained through *situatedness* (or being socially in place), and result in feelings about the place (or a sense-of-place). Kearns argued that the experience of medicine or care more broadly cannot be overtly detached from the place in which it is given and received, whether this be between different types of setting such as hospitals, community clinics or homecare or within those specific categories themselves. Since Kearns' call, a great deal of research has begun to conceptualize and understand places for health care not just as points or locations of human activities, but as a powerful social and cultural phenomenon (Kearns and Moon 2002). Although the vast majority of the new 'place-sensitive' health geography is focused on patients and consumption, following these disciplinary arguments and trends, part of a renewed engagement with health care has seen some attention to political economy, workers, work, and workplaces. These are considered briefly below.

As Andrews and Evans (2008) suggest, whilst interests in the distributive features of health care have remained particularly strong in the sub-discipline, since the early 1990s, moving beyond considerations of availability and use, studies have paid critical attention to the political and economic and processes that contribute to particular spatial patterns. This has included research on administrative and market boundaries (Cloutier-Fisher and Skinner 2006; Joseph and Chalmers 1996; Mohan 1998;), and policy and regulation (Joseph and Kearns 1996; Moon and Brown, 2000; Moon and North 2000; Norris 1997; Williams 2006). In combination, as Meade and Earickson (2000) outline, health care delivery systems have been recognized by geographers to exist in relation to social institutions, processes and beliefs, which has increased the attention paid by them to services for specific demographic and patient groups (Andrews et al. 2007; Peace et al. 2005).

In terms of workers, as Andrews and Evans (2008) outline, broad population-based studies within health geography have focused on the supply of labour, career decision-making, and the consequences of these for local communities. Location-based factors impacting on career choices, for example, have been found to include: educational experiences and clinical placements, preferences for urban living, wages, working conditions, work challenges, diversity of practice, career opportunities, career potential, family location, partner flexibility and preferences (Arminee and Nino 1996; Baer 2003; Cutchin 1997; Farmer et al. 2003; Guagliardo et al. 2004; Harrison 1995; Laditka 2004).

Andrews and Evans (2008) explain that with regard to work, certain studies consider the geographical dimensions and macro-scale consequences of decision-making across administrative bodies and/or professional groups (e.g., collective outcomes of health care administrators' decision making across service sites). These decisions are often narrowly defined in nature, involve specific financial, planning or clinical concerns, and are often reactions to policy changes (see Iredale et al. 2005; Moon et al. 2002; Twigg 1999). More recently there has been some attention to work processes and interactions, for example, related to hospital strategic management (Hanlon 2001), professional language (Gesler 1999), and the interpersonal and spatial dynamics that make and characterize specific work environments, including general/family practice (Rapport et al. 2006), neonatal intensive care (Brown and Middleton 2005), labour and delivery (Burges Watson et al. 2007), and mobile dialysis (Lehoux et al. 2007).

Regarding workplaces, in addition to some general attention to how facilities are perceived and used (Gesler and Kearns 2002), studies have focused on economic, structural and ideological/cultural changes (Kearns and Joseph 1997;

Kearns and Barnett 1997; Kearns et al. 2003; Prince et al. 2006). Specifically, at the system scale, they have considered the politics and symbolism of privatization and corporatization (Kearns and Barnett 1997) and, at the scale of single institutions, inter-professional conflicts over specific commercial activities (Kearns and Barnett 1999a, 1999b, 2000). Otherwise investigations into workplaces have centred on the emergence of new sites and settings in the broadening health care landscape. Whilst much attention here has been focused on home as a site of care, both at the level of regional planning (Williams 2006) and families and their dwellings (Dyck et al. 2005), other studies have focused on the emergence of complementary and alternative medicine and how its character is closely connected to its provision in orthodox medical settings, small private clinics and homes (Andrews 2004; Andrews and Phillips 2007).

As Andrews and Evans (2008) have suggested, the existence of this new 'post 1993' research demonstrates to critics that research on health care by geographers is not an abstract self serving spatial obsession, but is a body of work which makes valuable contributions to a wide-range of debates on health care. Moreover, it demonstrates that reform and place-sensitivity in health geography could and should move beyond consumption and include the study of production.

## Summary

As demonstrated in this chapter, geographical thoughts and inquiries on health care have been longstanding and diverse endeavors, ranging from the very earliest of medical ideas to contemporary developments in the social science of health geography. In many ways, the production of ideas has ebbed and flowed between disciplines. The health sciences, for example, have demonstrated particular leadership in classical literature and again in the age of exploration and most recently in the emergence of a geographical sub-discipline of nursing research. Health geography, meanwhile, has played a key role in developing the field in diverse ways in a comparatively short period of time since the mid-twentieth century. The current book must be read with this historical context in mind. The contributing authors have wide-ranging backgrounds, and through their chapters they each add both to their own disciplines' geographical traditions and to geographical perspectives on health care as a whole. Indeed, readers would do well to note their disciplinary contexts and backgrounds.

Informed by the observations in this chapter, and also by the chapters to follow, in Chapter 15 (the final chapter) we will make some recommendations about the future disciplinary constitution of geographical research on health care and specifically as it regards PHC. We will argue that although health geography does play a central role in the development of geographical perspectives on health care, given the continued presence and evolution of the health sciences in inquiry – particularly with regard to studying workplaces and work – as a discipline it must continue to refine its foci, and not move in a 'post-medical' direction that potentially excludes many ways of viewing and understanding, not just the delivery but also the practice of health care (see Andrews and Evans 2008). Moreover we highlight continued positive allied directions for the health sciences. All this is based on our committed view that it is possible to be simultaneously critical, applied, and relevant in the geographical study of health care and PHC.

## References

Andrews, G.J. (2004), '(Re)thinking the Dynamic Between Healthcare and Place: Therapeutic Geographies in Treatment and Care Practices', *Area* 36:3, 307–318.

Andrews, G.J. (2006), 'Geographies of Health in Nursing', *Health & Place* 12:1, 110–118.

Andrews, G.J. and Evans, J. (forthcoming), 'Understanding the Reproduction of Health Care: Towards Geographies in Health Care Work', *Progress in Human Geography*.

Andrews, G.J. and Moon, G. (2005), 'Space, Place and the Evidence Base, Part Two: Rereading Nursing Environment Through Geographical Research', *Worldviews on Evidence-Based Nursing* 2:3, 142–156.

Andrews, G.J. and Phillips, D.R. (2005), 'Petit Bourgeois Healthcare? The Big Small-business of Private Complementary Medical Practice', *Complementary Therapies in Clinical Practice* 11, 87–104.

Andrews, G.J. and Shaw, D. (2008), 'Clinical Geography: Nursing Practice and the Remaking of Institutional Space', *Journal of Nursing Management* 16:4, 463–473.

Andrews, G.J., Cutchin, M. and McCracken, K. et al. (2007), 'Geographical Gerontology: The Constitution of a Discipline', *Social Science & Medicine* 65:1, 151–168.

Andrews, G.J. Holmes, D. and Poland, B. et al. (2005), '"Airplanes are Flying Nursing Homes": Geographies in the Concepts and Locales of Gerontological Nursing', *International Journal of Older People Nursing* 14, 109–120.

Angus, J., Hodnett, E. and O'Brien-Pallas, L. (2003), 'Implementing Evidence-based Nursing Practice: A Tale of Two Intrapartum Nursing Units', *Nursing Inquiry* 10:4, 218–228.

Arminee, K. and Nico, P. (1996), 'Key Factors in Physicians' Choice of Practice Location: Findings from a Survey of Practitioners and their Spouses', *Health & Place*, 2:1, 27–34.

Baer, L.D. (2003), 'A Proposed Framework for Analysing the Potential Replacement of International Medical Graduates', *Health & Place* 9:4, 291–304.

Baron-Epel, O., Garty, N. and Green, M. (2007), 'Inequalities in Use of Health Services among Jews and Arabs in Israel', *Health Services Research* 42, 3, 1008–1019.

Baron-Epel, O., Kaplan, G. and Haviv-Messika, A. et al. (2007), 'Self-reported Health as a Cultural Health Determinant in Arab and Jewish Israelis MABAT – National Health and Nutrition Survey 1999–2001', *Social Science & Medicine* 61:6 1256–1266.

Bailey, T.C. and Gatrell, A.C. (1995), *Interactive Spatial Data Analysis* (Harlow: Longman).

Barrett. F. (2000), *Disease and Geography: The History of an Idea* (Toronto: Becker Associates).

Bender, A., Clune, L. and Guruge, S. (2007), 'Considering Place in Community Health Nursing', *Canadian Journal of Nursing Research* 39:3, 20–35.

Brawley, C.E. (2001), 'Environmental Design for Alzheimers Disease: A Quality of Life Issue', *Aging and Mental Health* 5:2, 79–83.

Brodie, D., Andrews, G.J. and Andrews, J.P. et al. (2005), 'Working in London Hospitals: Perceptions of Place in Nursing Students' Employment Considerations', *Social Science & Medicine* 61, 1867–1881.

Brown, S. and Middleton, D. (2005), 'The Baby as Virtual Object: Agency and Difference in a Neonatal Intensive Care Unit', *Environment and Planning D* 23, 695–715.

Burges-Watson, D., Murtagh. M.J. and Lally, J.E. et al. (2007), 'Flexible Therapeutic Landscapes of Labour and the Place of Pain Relief', *Health & Place* 13, 865–876.

Cackmak, S. (2007), 'Air Pollution: Uneven Distribution of Health Risks in Health Canada. People, Place and Health', *Health Policy Research Bulletin* 14, 33–37.

Carolan, M., Andrews, G.J. and Hodnett, E. (2006), 'Writing Place: A Comparison of Nursing Research and Health Geography', *Nursing Inquiry* 13:3, 203–219.

Cheek, J. (2004), 'Older People and Acute Care: A Matter of Place', *Illness, Crisis and Loss* 12:1, 52–62.

Cloutier-Fisher, D. and Skinner, M. (2006), 'Leveling the Playing Field? Exploring the Implications of Managed Competition for Voluntary Sector Providers of Long-term Care in Small Town Ontario', *Health & Place* 12:1, 97–109.

Creasey, J. (2007), 'Population Patterns in Canada', *Health Canada. Health Policy Research Bulletin* 14, 13–15.

Cutchin, M. (1997), 'Physician Retention in Rural Communities: The Perspectives of Experiential Place Integration', *Health & Place* 3:1, 25–41.

Delhorne, M. (1876), 'Projet d'une geographie medicale de la france, a l'useage des troups', *Journal de Medecine Militaire* 5, 137–153.

DesMeules, M., Luo, W. and Wang, F. (2007), 'Rural Places: Variations in Health', *Health Canada. Health Policy Research Bulletin* 14, 19–22.

Drake, D. (1850), *A Systematic Treatise Historical, Etiological and Practical on the Principal Diseases of the Interior Valley of North America as they Appear in the Caucasion, African, Indian and Esquimauz Varieties of its Population* (New York: Burt and Franklin).

Dyck, I., Kontos, P., Angus, J. and McKeever, P. (2005), 'The Home as a Site for Long Term Care: Meanings and Management of Bodies and Spaces', *Health & Place* 11:2, 173–185.

Eyles, J. (1990), 'How Significant are the Spatial Configurations of Health Care Systems?', *Social Science & Medicine* 31:1, 157–164.

Farmer J., Lauder W., Richards H. and Sharkey, S. (2003), 'Dr John has Gone: Assessing Health Professionals' Contribution to Remote Rural Community Sustainability in the UK', *Social Science & Medicine* 57, 4, 673–686.

Fell, D.B., Kephart, G. and Curtis, L. et al. (2007), 'The Relationship Between Work Hours and Utilization of General Practitioners in Four Canadian Provinces', *Health Services Research* 42:4, 1483–1498.

Finke, L.L. (1792–1795), *Versuch einer allgemeinen medicinisch-practischen geographie* (Bande Leipzig: Weidmann).

Fitzpatrick, J. and Whall, A. (1983), *Conceptual Models of Nursing: Analysis and Application* (London: Prentice Hall).

Gatrell, A.C. (2002), *Geographies of Health: An Introduction* (Oxford: Blackwell).

Gesler, W.M. (1999), 'Words in Wards', *Health & Place* 5:1, 13–25.

Gesler, W.M. and Kearns, R.A. (2002), *Culture/Place/Health* (London: Routledge).

Gilmour, J.A. (2006), 'Hybrid Space: Constituting the Hospital as a Home for Patients', *Nursing Inquiry* 13:1, 16–22.

Gould, P. and Leinbach, T.R. (1966), 'An Approach to the Geographic Assignment of Hospital Services', *Tijdschrift voor Economische en Sociale Geografie* 57, 203–206.

Gould, P. and Wallace, R. (1994), 'Spatial Structures and Scientific Paradoxes in the AIDs Pandemic', *Geografiska Annaler* 76b, 105–116.

Grenon, J., Butler, G. and Adams, R. (2007), 'Exploring the Intersection Between the Built Environment and Health Behaviors', *Health Canada. Health Policy Research Bulletin* 14, 29–32.

Guagliardo, M.F., Ronzio, C.R. and Cheung, I. et al. (2004), 'Physician Accessibility: An Urban Case Study of Pediatric Providers', *Health & Place* 10:3, 273–283.

Gupta, S. and Ross, N. (2007), 'Under the Microscope: Health Disparities Within Canadian Cities', *Health Canada Health Policy Research Bulletin* 14, 23–28.

Gupta, S. and Senzilet, L. (2007), 'Defining Measuring and Analysing Health and Place', *Health Canada. Health Policy Research Bulletin* 14, 10–12.

Halford, S. and Leonard, P. (2003), 'Space and Place in the Construction and Performance of Gendered Nursing Identities', *Journal of Advanced Nursing* 42:2, 201–208.

Halford, S. and Leonard, P. (2006a), *Negotiating Gendered Identities at Work: Place, Space and Time* (Basingstoke: Palgrave).

Halford, S. and Leonard, P. (2006b), 'Space, Place and Time: Contextualising Workplace Subjectivities', *Organizational Studies* 27:5, 657–676.

Hanlon, N. (2001), 'Sense of Place, Organizational Context and the Strategic Management of Publicly Funded Hospitals', *Health Policy* 58, 151–173.

Harrison, M.E. (1995), 'A Doctor's Place: Female Physicians in Mexico DF', *Health & Place* 1:2, 101–111.

Health Canada (2007), 'People, Place and Health', *Health Canada. Health Policy Research Bulletin* 14, 3–41.

Hodnett, E.D., Downe, S. Edwards, N. and Walsh, D. (2005), 'Home-like versus Conventional Institutional Settings for Birth (review)', *The Cochrane Database of Systematic Reviews* 1, 1–38.

Hoffmann, F. (1705), *Dissertatio morbis certis regionibus et populis propriis* (Halle: Henckilii).

Humboldt, A. (1807), *Ideen zu einer Geographie der Pflanzen nebst einem Naturgemalde der* (Tropenlander: Tubingen).

Iredale, R., Jones, L. Gray, J. and Deaville, J. (2005), 'The Edge Effect': An Explanatory Study of Some Factors Affecting Referrals to Cancer General Services in Rural Wales', *Health & Place* 11:3, 197–204.

Johnston, A.K. (1856), 'On the Geographical Distribution of Health and Disease, in Connection Chiefly with Natural Phenomenon', *The Physical Atlas of Natural Phenomenon*, 2nd edition.

Jones, K. and Moon, G. (1987), *Health, Disease and Society: An Introduction to Medical Geography* (London: Routledge).

Joseph, A. and Chalmers, A. (1996), 'Restructuring Long-term Care and the Geography of Ageing: A View from Rural New Zealand', *Social Science & Medicine* 42:6, 887–896.

Joseph, A.E. and Kearns, R.A. (1996), 'Deinstitutionalization Meets Restructuring: The Closure of a Psychiatric Hospital in New Zealand', *Health & Place* 2:3, 179–189.

Joseph, A.E. and Phillips, D.R. (1984), *Accessibility and Utilization: Geographical Perspectives on Health Care Delivery* (London: Harper & Row).

Kearns, R.A. (1993), 'Place and Health: Towards a Reformed Medical Geography', *The Professional Geographer* 46, 67–72.

Kearns, R.A. and Barnett, J. (1997), 'Consumerist Ideology and the Symbolic Landscapes of Private Medicine', *Health & Place* 3:3, 171–180.

Kearns, R.A. and Barnett, J. (1999a), 'Auckland's Starship Enterprise: Placing Metaphor in a Childrens' Hospital', in A. Williams, *Therapeutic Landscapes: The Dynamic Between Place and Wellness* (Lanham: University Press of America).

Kearns, R.A. and Barnett, J. (1999b), 'To Boldly Go? Place, Metaphor and Marketing of Auckland's Starship Hospital', *Environment and Planning D: Society and Space* 17, 201–226.

Kearns R.A. and Barnett, J.R. (2000), 'Happy Meals in the Starship Enterprise: Interpreting a Moral Geography of Food and Health Care Consumption', *Health & Place* 6:2, 81–93.

Kearns, R.A. and Joseph, A.E. (1997), 'Restructuring Health and Rural Communities in New Zealand', *Progress in Human Geography* 21:1, 18–32.

Kearns, R.A. and Moon, G. (2002), 'From Medical to Health Geography: Novelty, Place and Theory After a Decade of Change', *Progress in Human Geography* 26, 605–625.

Kearns, R.A., Barnett, J.R. and Newman, D. (2003), 'Placing Private Health Care: Reading Ascot Hospital in the Landscape of Contemporary Auckland', *Social Science & Medicine* 56:11, 2303–2315.

Kingma, M. (2006), *Nurses on the Move: Migration and the Global Health Care Economy* (London: Longwoods).

Kleffel, D. (1991), 'Rethinking the Environment as a Domain of Nursing Knowledge', *Advances in Nursing Science* 14:1, 40–51.

Laditka, J.N. (2004), 'Physician Supply, Physician Diversity and Outcomes of Primary Health Care for Older Persons in the United States', *Health & Place* 10:3, 231–244.

Lehoux, P., Poland, B. and Daudelin, G. et al. (2007), 'Designing a Better Place for Patients: Professional Struggles Surrounding Satellite and Mobile Dialysis Units', *Social Science & Medicine* 65:7, 1526–1548.

Liaschenko, J. (1994), 'The Moral Geography of Home Care', *Advances in Nursing Science* 17:2, 16–25.

Liaschenko, J. (1996), 'A Sense of Place for Patients: Living and Dying', *Home Care Provider* 1:5, 270–272.

Liaschenko, J. (1997), 'Ethics and the Geography of the Nurse-patient Relationship: Spatial Vulnerable and Gendered Space', *Scholarly Inquiry for Nursing Practice* 11:1, 45–59.

Liaschenko, J. (2003), 'At Home with Illness: Moral Understandings and Moral Geographies', *Ethica* 15:2, 71–82.

Lind, J. (1768), *An Essay on Disease Incidental to Europeans in Hot Climates, with the Method of Preventing their Fatal Consequences* (Becket: London).

Malone, R. (2003), 'Distal Nursing', *Social Science & Medicine* 56:11, 2317–2326.

McKeever, P., Stevens, B. and Miller, K.L. et al. (2002), 'Home verses Hospital Lactation Support for Newborns', *Birth* 29:4, 258–265.

Mohan, J. (1998), 'Explaining Geographies of Health Care: A Critique', *Health & Place* 4:2, 113–124.

Moon, G. and Brown, T. (2000), 'Governmentality and the Spatialized Discourse of Policy: The Consolidation of the Post-1989 NHS Reforms', *Transactions Institute British Geographers* 25:1, 65–76.

Moon, G. and North, N. (2000), *Policy and Place: General Practice in the UK* (Basingstoke: MacMillan).

Moon, G., Mohan, J. and Twigg, L. et al. (2002), 'Catching Waves: The Historical Geography of the General Practice Fundholding Initiative in England and Wales', *Social Science & Medicine* 55:12, 2201–2213.

Morrill, R.L. and Earickson, R. (1969), 'Location Efficiency in Chicago Hospitals', *Health Services Research* 4, 127–145.

Moss, M. and Schell, M. (2004), 'GISc A Scientific Framework and Methodological Tool for Nursing Research', *Advances in Nursing Science* 27:2, 150–159.

Nightingale, F. (1859), *Notes on Nursing: What it Is and What it is Not* (London: Lippincott).

Norris, P. (1997), 'The State and the Market: The Impact of Pharmacy Licensing on the Geographical Distribution of Pharmacies', *Health & Place* 3:4, 259–269.

Peace, S., Kellaher, L. and Holland, C. (2005), *Environment and Identity in Later Life* (New York: Open University Press).

Peter, E. (2002), 'The History of Nursing in the Home: Revealing the Significance of Place in the Expression of Moral Agency', *Nursing Inquiry* 9:2, 65–72.

Peter, E. and Liaschenko, J. (2004), 'Paradoxes of Proximity: A Spatio-temporal Analysis of Moral Distress and Moral Ambiguity', *Nursing Inquiry* 11:4, 218–225.

Prince, R., Kearns, R.A. and Craig, D. (2006), 'Govermentality, Discourse and Space in the New Zealand Health Care System', *Health & Place* 12:3, 253–266.

Purkis, M.E. (1996), 'Nursing in Quality Space: Technologies Governing Experiences of Care', *Nursing Inquiry* 3, 101–111.

Radcliffe, P. (1999), 'Geographical Mobility, Children and Career Progress in British Professional Nursing', *Journal of Advanced Nursing* 30:3, 758–768.

Rapport, F., Doel, M., Greaves, D. and Elwyn, G. (2006), 'From Manila to Monitor: Biographies of General Practitioner Workspaces', *Health* 10:2, 233–251.

Ritter, C. (1832), *Die Erdkunde in Verhaltbisse zur nature and zur Geschichte des Menschen: oder allgemeine Vergleichende geographie* (Berlin: Riemer).

Rosenberg, M.W. (1988), 'Linking the Geographical, the Medical and the Political in Analysing Health Care Delivery Systems', *Social Science & Medicine* 26, 179–186.

Secker, B., Goldberg, M. and Gibson, B. et al. (2006), 'Just Regionalization: Rehabilitating Care for People with Disabilities and Chronic Illness', *BMC Medical Ethics* 1–13.

Skelly, A., Arcury, T.A. and Gesler, W. et al. (2002), 'Sociospatial Knowledge Networks: Appraising Community as Place', *Research in Nursing and Health* 25, 159–170.

Smallman-Raynor, M., Muir, K.R. and Smith, S.J. (1998), 'The Geographical Assignment of Cancer Units: Patient Accessibility as an Optimal Location Problem', *Public Health* 112, 379–383.

Thorne, S., Canam, C. and Dahinten, S. et al. (1998), 'Nursing's Metaparadigm Concepts: Disimpacting the Debates', *Journal of Advanced Nursing* 27:6, 1257–1268.

Twigg, J. (1999), 'Regional Variation in Russian Medical Insurance: Lessons from Moscow and Nizhny Novgorod', *Health & Place* 5:3, 235–245.

West, E. and Barron, D.N. (2005), 'Social and Geographical Boundaries Around Senior Nurse and Physician Leaders: An Application of Social Network Analysis', *Canadian Journal of Nursing Research* 37:3, 132–148.

Williams, A. (2006), 'Restructuring Home Care in the 1990s: Geographical Differentiation in Ontario, Canada', *Health & Place* 12:2, 222–238.

# PART 1
## Practice and Delivery

# Chapter 3

# Access and Utilization Reconsidered: Towards a Broader Understanding of the Spatial Ordering of Primary Health Care

Neil Hanlon

Geographers have had a long engagement with primary health care (PHC), or at least primary care, as a focus for examining questions of welfare equity and distributive justice. Much of the more established work in PHC concentrates on measuring the spatial distribution of service providers and the impact of *friction of distance* (i.e., that overcoming distance requires extra time, effort, and material resources) on patterns of service utilization. Rather than review these developments, however, this chapter will argue that access and utilization studies need to pay greater attention to the full range of PHC options available to individuals in different settings (e.g., beyond physician-dominated primary care) and, as importantly, the structural issues and institutional dynamics at work in determining *who* provides PHC and *how* services are organized and distributed across space. PHC systems are important resources through which processes of social inclusion or marginalization are reinforced or contested. Broadening the conceptual and empirical scope of PHC geography promises to position the study of spatial access and utilization at the forefront not only of health research and policy evaluation, but of wider concerns of equity and social justice.

## Introduction

Access to PHC is a critical determinant of population health in any context, regardless of the level of wealth or social development. It is well known that having regular contact with PHC providers confers all sorts of benefits to individuals and the wider societies to which they belong. These benefits include more attention to health promotion, earlier detection of health problems, and more efficient use of other health sectors and resources, such as emergency services and tertiary level acute care (Macinko et al. 2007; Macinko et al. 2003; Starfield 1998). PHC refers not only to a broad collection of health practitioners (i.e., beyond physicians and nurses), but also a more holistic *approach* to health care that emphasizes health promotion and wellness (i.e., beyond the biomedical model). Thus, while questions abound about the contribution of the curative sectors of health systems to the overall health of populations, there is widespread support for policies that seek greater investments in PHC as a means to achieve better health and health care outcomes.

The realization of PHC's promise, however, depends upon the degree to which the full range of PHC resources available is engaged in the pursuit of population health objectives. There is often much emphasis on PHC as the point of initial contact in a given health system, but it should also be borne in mind that this sector plays an ongoing role in the patterning of care. This includes such things as the monitoring and management of chronic disease and degenerative conditions, rehabilitation from illness and injury, information and referral activities, and palliation. Consideration of the broader roles and activities of PHC leads us to think about a wider range of practitioners, other than physicians, who are engaged in the delivery of PHC. While the precise mix of practitioners varies both nationally and regionally, these might include fields such as community and public health nursing, pharmacy, dentistry, social work, occupational therapy and other allied professions. In fact, the rising popularity of various complementary and alternative (CAM) forms of treatment in many developed societies has generated much interest in whether and in what ways such fields as homeopathy, chiropractics and relaxation therapy might also be added to the mix (see Adams 2004; Del Casino 2004; Wahlberg 2007; Wiles and Rosenberg 2001).

Not surprisingly, there is widespread interest in the socio-spatial dimensions of PHC organization and delivery. Geographers are especially drawn to the fact that PHC is the aspect of formal health care systems where notions of distributional equity are most localized. That is, while no one would reasonably insist that specialists and sophisticated technologies be dispersed evenly across space, there is every reason to expect that PHC provision be distributed in a manner that is proportionate to the demographic and socio-economic structure of populations, but also sensitive to concerns of distance minimization for individuals in small and remote locations. Much of the more traditional work in PHC geography concentrates on relative and absolute measures of service distribution, or on the impact of friction of distance on patterns of utilization for various services at different areal, or geographically aggregated, levels of analysis. My interest here is not to summarize past and present work in these areas (see instead Cromley and McLafferty 2002; Gatrell 2002; Joseph and Phillips 1984; Ricketts et al. 1994), but rather to consider whether, and in what ways, we need to rethink our approach to access and utilization studies. One of the main objectives of the chapter is to develop the idea that the distribution of PHC resources over space, and the ways in which personnel and services are integrated in places, are central to the realization of PHC's capacity to achieve population health improvements. A second objective is to account for institutional barriers and filters through which health care systems are transformed. While there are many compelling forces working in favour of PHC development, the inertia of existing arrangements will have a major bearing on PHC reform and its ability to improve population health. In light of these two points, the chapter concludes with a call for multidimensional perspectives on access and utilization as critical benchmarks of the commitment to, and success of, PHC reform efforts.

## The Local Situatedness of Access and Utilization

At its core, access is the ability of an individual or group to obtain health services that they seek and that are available to them (see Kahn and Bhardwaj 1994 for a useful overview). Following the influential work of Aday and Andersen (1974), we can think of two basic forms of access: potential and realized. Potential access refers to the degree to which a population *might* make use of existing health resources, based on characteristics of available health services (e.g., supply and location) and of the populations who might use them (e.g., population size and demographic composition, exposure to risk factors). Realized access, on the other hand, is concerned solely with characterizing how a population *actually* makes use of health services available at a specified time period.

Geographers have long contributed to the literature on potential access by utilizing various indirect measures of a population's potential to use health resources in a specified service area, most often using municipalities, counties or districts because these are the spatial units for which the necessary data are readily available (e.g., Joseph and Phillips 1984; Ricketts et al. 1994). This work made use of various rates and ratios involving estimates of the quantity of health care resources available in the area as a numerator, and some estimate of the population or potential clients as a denominator. Measures such as location quotients estimate the share of resources relative to potential clients in a given area compared to the overall ratio of resources to clients (e.g., Anderson and Rosenberg 1990). The major advantages of these approaches are that reliable data are easy to obtain and the measures themselves are relatively straightforward to interpret. On the other hand, area estimates of health care personnel and potential clients are problematic (e.g., not all practitioners are equal in terms of qualifications and availability, not all residents of an area are equally likely to seek health care at any time). Second, the presumption that everyone seeks care within their county or municipality of residence is also problematic. In fact, the likelihood of so-called spillover or spatial by-passing is influenced by personal factors (e.g., age and socioeconomic status), ecological factors (e.g., geographic size and shape of jurisdictions), and the particular way in which populations and health resources are distributed within and between adjacent jurisdictions (e.g., Hanlon 2003; Hanlon and Skedgel 2006; Newton and Goldacre 1994; Radcliffe et al. 2003).

The application of spatial analysis techniques within a geographic information systems (GIS) framework has led to important breakthroughs in this spillover issue, more generally known as the modifiable area unit problem. For instance, recent work using floating catchment techniques based on kernel density estimations have resulted in far more sophisticated analysis of areas of resource scarcity (see Guagliano et al. 2004; Luo 2004; Wang 2007; Wang and Luo 2005). While this work has dealt exclusively with physicians, there are obvious benefits to be derived from applying these methods to analyses of a wider range of PHC personnel and resources. Better still, such techniques may hold great potential for examining potential access to various spatially contiguous bundles of PHC resources.

More recently, however, attention has shifted away from potential access and in the direction of the revealed form of access based on individual care seeking behaviour and experiences. This shift is the result of two otherwise disparate research developments. The first is the relaxed restrictions and greater availability of health care utilization data at the level of individual case records, including longitudinal files that have been linked to other population health databases and registries (e.g., Roos et al. 1998). This has resulted in a steady output of studies using increasingly sophisticated statistical and spatial analyses. Interestingly, this line of inquiry continues to draw heavily on Aday and Andersen's (1974) behavioural model of revealed access to characterize the extent to which health care utilization is influenced by various predisposing factors (e.g., age, health status, sex), enabling factors (e.g., income, knowledge of services, scope of available services) and barriers (e.g., price, travel costs, time constraints, language). Much of this work suggests that utilization is driven primarily by *need* (i.e., the presence of a health concern for which an individual seeks medical attention), but that those in need may face particular sets of barriers that vary by place and over time (e.g., Aday et al. 1980; Anderson and Newman 1973; Brameld and Holdman 2006; Finkelstein 2001; Iredale et al. 2005; Law et al. 2005). These barriers include such factors as cost, distance and the socio-cultural acceptability of available services (Brabyn and Barnett 2004; Dunlop et al. 2000; Rosenberg and Hanlon 1996; Wellstood et al. 2006).

The second development has been a rapid rise of qualitative inquiry examining the experiences of individuals who provide or obtain health care. This very broad research field draws upon work undertaken from disparate theoretical and conceptual standpoints, such as feminism, post-colonialism, queer theory, post-structuralism and hermeneutics (see Gesler and Kearns 2002). It would be inaccurate to characterize this body of work as a unified approach to access and utilization studies. Nor has there been any concerted effort to offer a new conceptual approach to access studies, as either a compliment or challenge to the Aday and Andersen framework. What these qualitative studies do share in common, however, is a commitment to understanding access and utilization from the perspectives of individuals negotiating various material and discursive forces acting upon them. This commitment is driven largely by an intellectual unease with statistical and spatially aggregated analyses that seek to reduce the unique experiences of individuals and local social networks to abstract analytical variables. Instead, qualitative inquiry seeks to highlight the particular situated challenges of individuals from vulnerable or marginalized groups such as international migrants, visible minorities and residents of remote areas (e.g., Asanin and Wilson 2008; Farmer et al. 2006; Gilgen 2005; Warfa 2006), to cite just a few examples.

Despite coming from very different, and in some cases irreconcilable, epistemological and methodological positions, the more recent turn towards direct approaches recognizes the importance of place and context in the delivery of PHC. This is in spite of major advances in communication technologies that have resulted in the fundamental restructuring of service delivery in other sectors such as personal finances, leisure and tourism. Overwhelmingly, the preferred mode of PHC delivery continues to be one in which providers and receivers are co-present. This is not to say that the delivery of PHC is fixed and unchanging. In fact, in spite of the developments highlighted above, the study of potential and revealed access must continue to play catch-up with the continually changing realities of PHC delivery. In the following section, I look more closely at different impetuses for change and transformation in the spatial organization and delivery of PHC, and consider what these might mean in terms of future patterns of access and utilization.

## Impetuses for the Spatial Reordering of PHC

As we move into the second decade of the new millennium, pressures delivering effective PHC are mounting on a number of fronts. For decades, reformers have called for a rebalancing of resource distribution between primary and tertiary levels of care, yet states continue to invest heavily in curative treatment technologies. While the *primary care* sector (i.e., general practitioners [GPs] and nurses) is often the main focus of policy attention in these rebalancing efforts, many jurisdictions face looming shortages of these practitioners (e.g., Canadian Institute for Health Information 2006). This fact, when coupled with varying degrees of population aging now affecting places around the globe, suggests strongly that there will be ever greater time pressures on GPs that will likely result in reduced access to care and less intensive use of their services in the future. Such a scenario underscores the need not only to increase the supply of GPs, but also to find new ways to deliver *PHC* through the use of various allied health personnel in a more coordinated and seamless system.

Not surprisingly, there have been growing pressures to transform the delivery of primary care as a necessary step towards PHC reform. One means of doing this has been to encourage GPs to abandon the tradition of solo practice in favour of more collaborative and interdisciplinary environments. The most basic of these transformations involve GPs sharing patient rosters, rather than simply sharing the same office space or clinic setting. Roster-sharing enables greater choice and convenience for patients, but more importantly it makes possible greater collaboration in dealing with complex cases, scheduling backlogs, and informational continuity of care in cases where individual GPs leave a practice location.

A more fundamental change in practice, however, would see a move away from physician-dominated primary care and towards more multidisciplinary PHC delivery. This would have GPs working in greater collaboration with various allied health professionals, such as nurse practitioners, pharmacists, dentists, ophthalmologists and occupational therapists (e.g., Solheim et al. 2007). Such transformation remains elusive, despite efforts to promote PHC teams in various national settings (e.g., Few et al. 2003; Jesson and Wilson 2003; Parfitt and Cornish 2007; Reeves and Baker 2004). In Canada, for instance, the Federal Government committed \$150 million in the late 1990s to pilot projects that promised innovative and integrated PHC delivery, including collaboration among various health professional bodies and health sectors. In all, 140 projects were established, including 93 that involved some aspect of multidisciplinary collaboration between GPs and other health professionals (Health Canada, no date). This was followed by the \$800 million Primary Health Care Transition Fund (2000–2006) in support of PHC reform efforts. Notwithstanding these initiatives, a more fundamental 'transition' from solo practice and physician-dominated primary care to a collaborative and integrated PHC has yet to take hold in Canada (e.g., Hutchison et al. 2001).

A third impetus for the transformation of PHC is the decreasing friction of distance affecting health care seeking behaviour. This is part of a long-term trend that has underpinned the consolidation and centralization of health care service provision, including primary level services. In fact, populations have become increasingly sub- and ex-urbanized, often ahead of corresponding redistributions of services and resources. Rising rates of automobile ownership and use make it possible for individuals to travel further to obtain PHC, just as urban sprawl makes this health care seeking behaviour more necessary. This type of spatial behaviour has, in some instances, even been encouraged in various policy approaches that promote greater choice for patients and more internal competition amongst providers (see Exworthy and Peckham 2006). This is not to suggest, however, that all members of society are equally willing and able to overcome greater distances to obtain PHC. For many socially marginalized groups (e.g., single mothers, persons with disabilities, seniors on fixed incomes, refugees and internally displaced persons), distance and time remain important constraints on health care seeking behaviour (see Yanos 2007; Young 1999). But there is clear evidence that a substantial share of the population in developed societies are willing and able to travel further for health care (e.g., Bramdel and Holman 2007; Hanlon 2003; Hanlon and Skedgel 2006; Iredale et al. 2005). Such behaviour poses a challenge to some of the long-standing assumptions about access on which many health systems were planned, including several of the traditionally reliable measures and methods of examining potential access (Powell and Exworthy 2003).

Finally, the growing interest in, and more importantly use of, so-called CAM must certainly be regarded as a development that will potentially transform the organization and distribution of PHC (Adams 2004; Wahlberg 2007). CAM treatments are now a permanent feature of the health service landscape. Together with calls to allow nurse practitioners, pharmacists and dentists in particular to play greater roles in PHC delivery, the study of access and utilization will have to accommodate the emergence of these forms of care, including their spatial patterning and relationship with established physician services (e.g., as a complement to or substitution for GP services).

These impetuses for change promise to continue to transform the organization and spatial distribution of PHC in important ways. Such developments point to the need for a much wider engagement with emerging forms of PHC delivery in the health geography literature. While some time has now passed since Kearns' (1993) call for a post-medical geography, geographers remain somewhat slow off the mark when it comes to engaging with a wider array of primary care practitioners. That is, despite some notable exceptions (e.g., Conradson 2003; Del Casino 2004; Johnsen et al. 2005; Wiles and Rosenberg 2001), health geographers have tended to remain fixated on GPs as a critical marker of PHC access and utilization. Such an approach overlooks the scope and potential for people to arrange PHC within the constraints of availability, affordability and acceptability.

## Taking Account of Institutional Forces Acting on PHC Development

For all of the forces acting in favour of PHC reform, there is much about existing health care arrangements that poses a significant barrier to change. Analysts of health care policy and reform have employed the conceptual framework of *new institutionalism* to make sense of the inherent conservatism of health care systems. Institutionalism has had considerable influence in various fields of political economy over the past three decades, including other areas of human geography research (see Amin 2001; Martin 2000). To date, however, the literature on health care access and utilization has yet to engage with these lines of inquiry, especially as it pertains to questions of reform centred on PHC development.

Institutionalism seeks to account for the role of socially-embedded routines, norms and conventions in shaping the operation of economic and political systems (Immergut 1998). The approach is concerned not only with organizational arrangements and various activity structures, but also the very nature of interpersonal relations and behaviours that construct, reproduce or transform social systems. In its broadest sense, therefore, institutions are 'cognitive, normative and regulative structures and activities that provide stability and meaning to social behaviour' (Scott 1995, 33).

Institutional arrangements are the outcomes of historically-situated negotiations amongst asymmetrically related networks of actors and groups. The inertia of existing arrangements, which often entails the accommodation of various interest groups, poses a barrier to change even as the social conditions underpinning these arrangements become less relevant or compelling. Institutional perspectives suggest that change in these political accommodations is only likely to occur incrementally until such time as institutional features appear to become outmoded or ill-adapted to changing exogenous conditions. It is only at this point that these accommodations become vulnerable to major upheaval. An institutional perspective, however, also anticipates substantial limitations on the capacity of reforms and reformers to replace or otherwise transform existing arrangements of status or privilege. Many proponents of PHC reform have argued that undertaking institutional change is not only needed but also past due.

A key feature of health care policy, particularly in industrialized countries, is its path dependency (i.e., the continued reliance on historical precedents to shape contemporary health care policy and practice) (Tuohy 1999). This is not to say that major reform initiatives are not undertaken, nor that these efforts are necessarily destined to fall well short of their objectives. Indeed, there has been a spate of reform efforts looking to achieve major changes in the organization and delivery of health care around a variety of issues, including cost control, decentralized planning and decision making, de-institutionalization, patient-centredness, and a greater role for primary care, if not PHC. It is well beyond the scope and intent of this chapter to catalogue these reform efforts, or provide a general summary; suffice to say that a common refrain in the literature is that the implementation of reform initiatives rarely assumes the form originally intended (e.g., Brabyn and Barnett 2004; Naylor 1999; Moran 1999; Powell and Exworthy 2003).

Many of the institutional arrangements made to accommodate greater state intervention in health insurance in Canada in the 1950s and 1960s, for instance, remain the cornerstone of provincially administered 'medicare' programs today. GPs continue to be the sole gatekeepers of referral to specialists and retain a near monopoly on the authority to prescribe pharmaceuticals. Likewise, physicians retain considerable control of medical education, professional licensure and location of practice. Governments and private insurers remain risk averse when it comes to empowering other health professionals, such as nurse practitioners, community health nurses and pharmacists to provide PHC. The anticipated power struggle with an organized and well funded medical profession has proved a formidable obstacle to any efforts to broaden the scope of PHC responsibilities amongst various allied health professions. These same considerations have been cited for the lengthy process of replacing fee-for-service payment with alternatives such as capitation in Great Britain (e.g., Tuohy 1999), and the persistence of fee-for-service remuneration in Australia and Canada (e.g., Hall 1999; Shortt 1999).

In summary, an institutional perspective on the issue of barriers to PHC development emphasizes the innate conservatism of existing arrangements. The discussion above highlights the medical profession as a powerful vested interest in the health care politics of developed societies. Accommodations that privilege certain health care professionals over others will tend to fuel the professional aspirations of marginalized provider groups. There is a burgeoning literature in nursing that describes the increasing creep of medicalization as a consequence of professionalization, but also of ongoing health care reforms and restructuring (e.g., Malone 2003). Some worry that these tendencies may also impact the practice of allied care and CAM as these become more mainstream features of PHC systems (e.g., Bondi 2004; Jesson and Wilson 2003). These and other institutional pressures will continue to exert considerable influence over who provides PHC and how these services are delivered, with all that this entails for future patterns of access and utilization. Such developments point to the ongoing need for in-depth examinations of PHC delivery to add meaning to the maps and statistics summarizing the distribution of personnel and utilization of services.

## Conclusion

PHC is widely regarded to be a major feature of the population health framework for health reform. Having regular contact with primary-level practitioners has been shown to offer more effective illness prevention, earlier diagnosis and treatment of disease, more integrated follow-up care and referrals to specialists, and greater uptake of advice and informational support with health promotion. By engaging a wider range of practitioners and encouraging greater cooperation and collaboration amongst professional groups, PHC offers a great deal of flexibility in the delivery of health care, including opportunities to address longstanding health care deficiencies in socially and spatially disadvantages areas. For these reasons, investment in PHC development can be viewed as a major expression of a society's commitment to social inclusion and community development.

In spite of this promise and potential, however, reform based on PHC principles has tended to be characterized as slow and incremental. This has prompted concerns as to what degree medically dominated health systems can accommodate major reform along lines such as PHC reorientation (e.g., Hutchison et al. 2001). In light of this, perhaps it is not surprising that the health geography literature on access and utilization has retained its emphasis on primary care rather than the wider concerns of PHC. And yet there is great potential for geographers to position the study of access and utilization at the forefront of PHC reform research and evaluation.

In this chapter, I have highlighted some recent developments in the study of potential and revealed or realized access that hold much promise for broader applications in the full range of PHC delivery. I have also highlighted a lack of attention to date with institutional forces known to dampen, prevent or otherwise reshape reform initiatives by influencing the content and spatial arrangement of PHC systems. Here there is an opportunity for future research to fill an important role by conducting in-depth and conceptually grounded analyses of PHC delivery that exposes the tensions and fissures between medical and population health orientations. Together, these developments point to a major role for access and utilization research in measuring the level of equity in the spatial distribution of PHC resources, and the degree to which these resources are effectively integrated in places as a means of achieving population health objectives.

Extending health geography's engagement with primary care to a focus on PHC may reinvigorate the study of access and utilization, but this will almost certainly continue to be a pluralist rather than unified effort. At present, those engaged in spatial analysis, epidemiological approaches and qualitative inquiry do little more than communicate across methodological and disciplinary divides at conferences and in published works, as opposed to engaging in truly multidisciplinary and interdisciplinary efforts to address these issues. This may well continue to be the case, given the very disparate and often opposing intellectual standpoints between, but also at times within, these groups. Certainly the era of general theory construction in social science research is now long over, and there seems little interest in, or appetite for, revisiting past efforts to derive one-size-fits-all definitions and measures of access. Multidimensional perspectives on access and utilization that come from a variety of research approaches appear to offer the best means of moving forward. While access and utilization are appropriately viewed as critical benchmarks of a commitment to, and the success of, PHC reform efforts, the diverse forms that PHC takes, the complex interplay of individual and social factors and, finally, the institutional aspects of health care delivery all point to a need to continue to encourage a diversity of access and utilization research within and beyond health geography.

## References

Adams, J. (2004), 'View Point: Exploring the Interface Between Complementary and Alternative Medicine (CAM) and Rural General Practice: A Call for Research', *Health & Place* 10:3, 285–7.

Aday, L. and Andersen, R. (1974), 'A Framework for the Study of Access to Medical Care', *Health Services Research* 2, 208–20.

Aday, L., Andersen, R. and Fleming, G. (1980), *Health Care in the U.S.: Equitable for Whom?* (Beverly Hills: Sage).

Amin, A. (2001), 'Moving On: Institutionalism in Economic Geography', *Environment and Planning A* 33, 1237–41.

Andersen, R. and Newman, J. (1973), 'Societal and Individual Determinants of Medical Care Utilization in the United States', *The Milbank Memorial Fund Quarterly* 51, 95–124.

Anderson, M. and Rosenberg, M. (1990), 'Ontario's Underserviced Area Program Revisited: An Indirect Analysis', *Social Science & Medicine* 30:1, 35–44.

Asanin, J. and Wilson, K. (2008), '"I Spent Nine Years Looking for a Doctor": Exploring Access to Health Care Among Immigrants in Mississauga, Ontario, Canada', *Social Science & Medicine* 66:6, 1271–83.

Bondi, L. (2004), '"A Double-edged Sword?" The Professionalisation of Counselling in the United Kingdom', *Health & Place* 10:4, 319–28.

Brabyn, L. and Barnett, J. (2004), 'Population Need and Geographic Access to General Practitioners in Rural New Zealand', *New Zealand Medical Journal* 117, 1199.

Brameld, K. and Holman, C.D.J. (2006), 'The Effect of Locational Disadvantage on Hospital Utilization and Outcomes in Western Australia', *Health & Place* 12:4, 490–502.

Canadian Institute for Health Information (2006), *Health Personnel Trends in Canada, 1995 to 2004* (Ottawa: The Institute).

Conradson, D. (2003), 'Spaces of Care in the City: The Place of a Community Drop-in Centre', *Social and Cultural Geography* 4:4, 507–25.

Cromley, E. and McLafferty, S. (2002), *GIS and Public Health* (New York and London: Guilford Press).

Del Casino Jr., V. (2004), '(Re)placing Health and Health Care: Mapping the Competing Discourses and Practices of "Traditional" and "Modern" Thai Medicine', *Health & Place* 10:1, 59–73.

Dunlop, S., Coyte, P. and McIsaac, W. (2000), 'Socio-economic Status and the Utilization of Physicians' Services: Results from the Canadian National Population Health Survey', *Social Science & Medicine* 51:1, 123–33.

Exworthy, M. and Peckham, S. (2006), 'Access, Choice and Travel: Implications for Health Policy', *Social Policy and Administration* 40:3, 267–87.

Farmer, J., Iverson, L. and Campbell, N. et al. (2006), 'Rural/Urban Differences in Accounts of Patients' Initial Decisions to Consult Primary Care', *Health & Place* 12:2, 210–21.

Few, R., Harpham, T. and Atkinson, S. (2003), 'Urban Primary Health Care in Africa: A Comparative Analysis of City-wide Public Sector Projects in Lusaka and Dar es Salaam', *Health & Place* 9:1, 45–53.

Finkelstein, M. (2001), 'Do Factors Other than Need Determine Utilization of Physicians' Services in Ontario?', *Canadian Medical Association Journal* 165:5, 565–70.

Gatrell, T. (2002), *Geographies of Health: An Introduction* (Oxford: Blackwell).

Gesler, W. and Kearns, R. (2002), *Culture/Place/Health* (London and New York: Routledge).

Gilgen, P., Maeusezahl, D. and Salis Gross, C. et al. (2005), 'Impact of Migration on Illness Experience and Help-seeking Strategies of Patients from Turkey and Bosnia in Primary Health Care in Basel', *Health & Place* 11:3, 261–73.

Guagliardo, M., Ronzio, C. and Cheung, I. et al. (2004), 'Physician Accessibility: An Urban Case Study of Pediatric Providers', *Health & Place* 10:3, 273–83.

Hall, J. (1999), 'Incremental Change in the Australian Health Care System', *Health Affairs* 18:3, 95–110.

Hanlon, N. (2003), 'Measuring Aspects of Devolved Health Authority Performance: Nova Scotia Patients Who Travel Further than Necessary to Obtain Hospital Care', *Healthcare Management Forum* 16, 8–13.

Hanlon, N. and Skedgel, C. (2006), 'Cross-district Utilization of General Hospital Care in Nova Scotia: Policy and Service Delivery Implications for Rural Districts', *Social Science & Medicine* 62:1, 145–56.

Hutchison, B., Abelson, J. and Lavis, J. (2001), 'Primary Care in Canada: So Much Innovation, So Little Change', *Health Affairs* 20:3, 116–31.

Immergut, E. (1998), 'The Theoretical Core of the New Institutionalism', *Politics and Society* 26, 5–34.

Iredale, R., Jones, L., Gray, J. and Deaville, J. (2005), 'The Edge Effect': An Exploratory Study of Some Factors Affecting Referrals to Cancer Services in Rural Wales', *Health & Place* 11:3, 197–204.

Jesson, J. and Wilson, K. (2003), 'One-stop Health Centres: What Co-location Means for Pharmacy', *Health & Place* 9:3, 253–61.

Johnsen, S., Cloke, P. and May, J. (2005), 'Transitory Spaces of Care: Serving Homeless People on the Street', *Health & Place* 11:4, 323–36.

Joseph, A. and Phillips, D. (1984), *Accessibility and Utilization: Geographical Perspectives on Health Care Delivery* (New York: Harper & Row).

Kahn, A. and Bhardwaj, S. (1994), 'Access to Health Care', *Evaluation and the Health Professions* 17:1, 60–76.

Kearns, R. (1993), 'Place and Health: Toward a Reformed Medical Geography', *The Professional Geographer* 45, 139–47.

Law, M., Wilson, K. and Eyles, J. et al. (2005), 'Meeting Health Need, Accessing Health Care: The Role of Neighbourhood', *Health & Place* 11:4, 367–77.

Luo, W. (2004), 'Using a GIS-based Floating Catchment Method to Assess Areas with Shortages of Physicians', *Health & Place* 10:1, 1–11.

Macinko, J., de Souza, M., Guanais, F. and Simoes, C. (2007), 'Going to Scale with Community-based Primary Care: An Analysis of the Family Health Program and Infant Mortality in Brazil, 1999–2004', *Social Science & Medicine* 65, 2070–80.

Macinko, J., Starfield, B. and Shi, L. (2003), 'The Contribution of Primary Care Systems to Health Outcomes within Organization for Economic Cooperation and Development (OECD) Countries, 1970–1998', *Health Services Research* 38, 831–865.

Malone, R. (2003), 'Distal Nursing', *Social Science & Medicine* 56:11, 2317–26.

Martin, R. (2000), 'Institutional Approaches in Economic Geography', in Sheppard, E. and Barnes, T. (eds), *A Companion to Economic Geography* (Malden: Blackwell), 77–94.

Moran, M. (1999), *Governing the Health Care State: A Comparative Study of the United Kingdom, the United States and Germany* (Manchester and New York: Manchester University Press).

Naylor, C.D. (1999), 'Health Care in Canada: Incrementalism Under Fiscal Duress', *Health Affairs* 18, 9–26.

Newton, J. and Goldacre, M. (1994), 'How Many Patients are Admitted in Districts Other than Their Own, and Why?', *Journal of Public Health Medicine* 16, 159–64.

Parfitt, B. and Cornish, F. (2007), 'Implementing Family Health Nursing in Tajikistan: From Policy to Practice in Primary Health Care Reform', *Social Science & Medicine* 65, 1720–29.

Powell, M. and Exworthy, M. (2003), 'Equal Access to Health Care and the British National Health Service', *Policy Studies* 24:1, 51–64.

Radcliff, T., Brasure, M., Moscovice, I. and Stensland, J. (2003), 'Understanding Rural Hospital Bypass Behavior', *Journal of Rural Health* 19:3, 252–59.

Reeves, D. and Baker, D. (2004), 'Investigating Relationships Between Health Need, Primary Care and Social Care Using Routine Statistics', *Health & Place* 10:2, 129–40.

Ricketts, T., Savitz, L., Gesler, W. and Osborne, D. (eds) (1994), *Geographic Methods for Health Services Research: A Focus on the Rural–Urban Continuum* (New York: University Press of America).

Roos, N., Black, C. and Roos, L. et al. (1998), 'Managing Health Services: How Administrative Data and Population-based Analysis Can Focus the Agenda', *Health Services Management Research* 11, 49–67.

Rosenberg, M. and Hanlon, N. (1996), 'Access and Utilization: A Continuum of Health Service Environments', *Social Science & Medicine* 43:6, 975–83.

Scott, W. (1995), *Institutions and Organizations* (Thousand Oaks: Sage).

Shortt, S. (1999), *The Doctor Dilemma: Public Policy and the Changing Role of Physicians Under Ontario Medicare* (Montreal and Kingston: McGill-Queen's University Press).

Solheim, K., McElmurry, B. and Kim, M. (2007), 'Multidisciplinary Teamwork in US Primary Health Care', *Social Science & Medicine* 65, 622–34.

Starfield, B. (1999), *Primary Care: Balancing Health Needs, Services, and Technology* (New York and Oxford: Oxford University Press).

Tuohy, C. (1999), *Accidental Logics: The Dynamics of Change in the Health Care Arena in the United States, Britain and Canada* (Oxford and New York: Oxford University Press).

Wahlberg, A. (2007), 'A Quackery with a Difference – New Medical Pluralism and the Problem of 'Dangerous Practitioners' in the United Kingdom', *Social Science & Medicine* 65, 2307–16.

Wang, F. and Luo, W. (2005), 'Assessing Spatial and Non-spatial Factors for Healthcare Access: Towards an Integrated Approach to Defining Health Professional Shortage Areas', *Health & Place* 11:2, 131–46.

Wang, L. (2007), 'Immigration, Ethnicity, and Accessibility to Culturally Diverse Family Physicians', *Health & Place* 13:3, 656–71.

Warfa, N., Kamaldeep, B. and Craig, T. et al. (2006), 'Post-migration Geographical Mobility: Mental Health and Health Service Utilization Among Somali Refugees in the U.K.: A Qualitative Study', *Health & Place* 12:4, 503–15.

Wellstood, K., Wilson, K. and Eyles, J. (2006), '"Reasonable Access" to Primary Care: Assessing the Role of Individual and System Characteristics', *Health & Place* 12, 121–130.

Wiles, J. and Rosenberg, M. (2001), '"Gentle Caring Experience" – Seeking Alternative Health Care in Canada', *Health & Place* 7:3, 209–24.

Yanos, P.T. (2007), 'Beyond "Landscapes of Despair": The Need for New Research on the Urban Environment, Sprawl, and Community Integration of Persons with Severe Mental Illness', *Health & Place* 13:3, 672–76.

Young, R. (1999), 'Prioritising Family Health Needs: A Time-space Analysis of Women's Health-related Behaviours', *Social Science & Medicine* 48, 797–813.

**Internet-based References**

Health Canada (n.d.), 'Health Transition Fund'. Available at http://www.hc-sc.gc.
    ca/index_e.html. Accessed April 1, 2008.

Chapter 4

# The Effects of Population Density, Physical Distance and Socio-Economic Vulnerability on Access to Primary Health Care in Rural and Remote British Columbia, Canada

Nadine Schuurman

Rural residents of Canada and many other countries often lack equitable access to primary health care services. Physical distance remains the greatest impediment to providing primary health care in rural, remote and northern areas. It is likely that not only rurality but also socio-economic status (SES) affects health outcomes. This chapter examines both influences on access to primary health care generally as well as to specific health services. The aim is to describe the potential for using geographic information systems (GIS) to: (1) define rational catchments based on travel to primary health care services; (2) characterize levels of access to certain forms of primary health care in the Canadian province of British Columbia (BC); (3) describe the extent of under-service in non-urban BC for selected services; and (4) explicate the impact of combined social economic vulnerability and distance to services on rural residents. A case study of the Canadian province of BC is used to achieve these aims. Maps and figures are used to dramatize and communicate the effects of distance and social vulnerability on rural residents.

In the 1960s universally accessible health care was introduced in Canada. Since then (and increasingly from the early 1990s on) sustainability has preoccupied federal and provincial governments in Canada, ultimately leading to attempts at health care reform and restructuring in the form of regionalization in most provinces (see Church and Barker 1998). The primary aim of health care restructuring is the rationalization of services to contain costs. Hospital budgets are commonly the first place that governments begin to reduce expenses (Barer et al. 2003). Regionalization, involving both downloading of authority and responsibility from provincial jurisdiction to regional bodies along with a parallel upward movement of authority and responsibility from more local administrative bodies, is another restructuring outcome. As a result, rural and remote areas are frequently subject to service rationalization in which primary health care services, including maternity and trauma care, in smaller centres are closed and larger centres – often considerably more distant – provide services to a larger catchment. The emphasis on rationalization of hospital services has a dramatic effect on rural and remote populations as their access to some forms of primary health care (and medical care in particular) is frequently through local hospitals, particularly in communities that do not have on-site primary care clinics.

Analysing access to health care services is a growing pursuit as fiscal pressure increases. Health managers want maximum return for health expenditures – with an attendant focus on fiscal rationalization. Access can be measured using three indicators: client needs (e.g., are clients able to get needed services?), affordability (e.g, are clients able to afford services?) and client acceptability (e.g. is the state of service accessibility acceptable to clients?). In assessing each of these, there is an emphasis on service delivery systems. Not coincidentally, location and allocation models typically used in marketing are also used for health decision support systems (Cromley and McLafferty 2002). The goal in both cases is to optimize service for the lowest possible cost.

Hospitals are an important site of certain forms of primary health care provision, though not always considered as such. Specifically, such services delivered in hospitals include maternity care and trauma care; they are essential services and thus central to the vision of providing health for all through primary health care. Watson et al. (2005), for example, include visits to hospital emergency rooms (i.e., the site where trauma care is delivered from) in their evaluation of primary health care services in BC and maternity care is explicitly mentioned in the Declaration of Alma-Ata in principle VII.c (see Chapter 1). Equity of access to 'first contact' health services is at the core of primary health care, and is also an explicit goal of the BC provincial government, including with regard to maternity and trauma care. Equity of access is also directly linked to health outcomes. To that end, this chapter examines the effect that hospital and service closures have on rural populations.

I begin by describing rural vulnerability to poorer health outcomes, the role of socio-economic status on health, and the importance of catchments or health service areas to determine access to services. The second section of the chapter uses GIS to illustrate the extent to which rural populations in BC are served by primary health care services, using rural maternity and access to trauma services as examples. I illustrate the effect of service closures on catchment size and population, and the degree to which socio-economic status may exacerbate vulnerability.

**Rural Vulnerability**

Geography has been recognized as a determinant of health in Canada, with rural Canadian populations experiencing poorer health than urban Canadians (Furuseth 1998; Pong and Pitblado 2003; Romanow 2002). Canadian life expectancy is among the world's highest at 79 years, but is geographically uneven, falling below 75 years in some rural communities (Furuseth 1998). Rural areas tend to be undersupplied by physicians with the outcome that seeing a physician at all can be challenging. Although this is a different issue than accessibility, it remains a factor in rural health vulnerability (Joseph and Bantock 1982). Poorer rural health outcomes have been attributed to the relative inaccessibility of quality primary health care services and health resources for rural Canadians (Romanow 2002; Ryan-Nicholls 2004; Sutherns et al. 2003).

Unequal distribution of access to primary health care and health care resources negatively affects rural populations across the world – as well as those in Canada. Medical and health geographers have struggled with methods to characterize this inequality of access including using statistics to characterize inadequate coverage (e.g., graphing expenditures versus need), and mapping areas that have lower than average physician-to-patient ratios (Meade and Erickson 2000). Other health geographers have examined socio-cultural and political influences that affect the delivery of health care (e.g., Rosenberg 1988). Many of these methods account for distance from services and the close link between access and health outcomes.

Access to health services – including primary health care – is one of the most important factors in determining population health (Guagliardo 2004). Not only do rural residents have the least choice of services, policies tend to favour the urban majority, especially with respect to resource allocation (Schuurman et al. 2006). Studies confirm rural vulnerability with respect to health and suggest that rural culture supports behaviours that may have negative health effects (Cromley and McLafferty 2002; Hartley 2004). In the United States (US), rural residents have been found to smoke more while exercising less and demonstrating higher rates of obesity (Hartley 2004). Mortality rates for motor vehicle crashes, asthma and cancer are all higher in rural areas (Baird and Wright 2006). A US study found that numbers of motor vehicle crashes (MVC) were significantly higher among rural residents of the state of Arizona (Campos-Outcalt et al. 2003). Certainly, MVC survival rates are also linked to rurality – among other factors (Miles-Doan and Kelly 1995). Non-metropolitan residence in the US is associated with a higher number of low birth weight babies. Moreover it is associated with later access to prenatal care as well as a greater number of adverse outcomes (Larson et al. 1997). Likewise, inequalities in access to and provision of services exist for rural populations in the United Kingdom (UK) (Swindlehurst et al. 2005). Rural patients in UK are admitted to hospital less frequently than their urban counterparts and suffer poorer health outcomes (Baird and Wright 2006). Each of these studies confirms that the burden of disease is higher amongst rural residents, with a clear link between access to care and outcome.

Driving time or geographic access is especially relevant in examining health outcomes in rural populations. One study in the US found that 60 percent of rural residents chose the closest possible hospital for accessing services. However, when the number of beds in the hospital was under 75, only 54 percent accessed services at the smaller, closer hospital. By contrast, 79 percent of rural residents utilized the closest services when the hospital size was over 75 beds. These findings have implications for concentration of services in health planning, as patients try to balance shorter driving times with higher levels of service (Adams and Wright 1991). A detailed survey of over 1,000 residents in the US state of North Carolina found that distance to care and having a driver's license influenced outcomes

(Arcury et al. 2005a). A subsequent analysis found that having a driver's license resulted in 2.29 times more health care visits for chronic care and 1.92 times more visits for regular check-up care compared to rural residents without a license. Likewise, rural residents with close family or friends who were willing to drive them to care visited health care services 1.58 times more often than more isolated residents (Arcury et al. 2005b).

There are, however, some researchers who disagree with the axiom that health inequality is associated with rurality. Liu et al. (2001) conducted a study in the Canadian province of Saskatchewan. The study compared communities that had experienced hospital closure with three groups: (1) rural communities that never had a hospital; (2) rural communities that still have a hospital; and (3) the rest of Saskatchewan. The authors conclude that rural hospital closures (in Saskatchewan) did not adversely impact rural residents' health status, but their view was tempered with an understanding that this conclusion was not entirely consistent with the perceptions of residents in that they were generally opposed to closures (Liu et al. 2001). Closures such as those in Saskatchewan place the burden of distance on patients regardless of the impact on health outcomes.

Rurality is one factor affecting health outcomes and longevity but it is often compounded by local and individuals socio-economic factors. The relationship between SES and ill health is well-documented (Faelker et al. 2006; Gilthorpe et al. 1997; Orpana and Lemyre 2004; Sin et al. 2001; Veenstra et al. 2005). SES has been shown to be a much more powerful agent than genetics in determining health outcomes (Baird and Wright 2006).

Access and provision issues extend to aboriginal populations (Wardman et al. 2005). A study of palliative care amongst First Nations in rural areas of the Canadian province of Manitoba developed several policy guidelines including that patient-centered care plans be developed for rural and remote communities and better communication be established between rural practitioners and urban specialists (Hotson et al. 2004). Issues related to palliative care in rural and remote communities are common internationally. An Australian study found that generalist rural doctors not only treat many dying patients, but also provide a continuity of care that is rarely seen in other settings (Pereira 2005). These factors must all be integrated into a vision of rural health vulnerability and mitigating socio-economic factors.

## Catchments and Access to Rural Primary Health Care Services

A catchment is defined as a geographical area delineated around an institution or business that includes the client population that uses its services. Catchments normally divide the landscape into contiguous regions, but they can also be overlapping. Accurate catchment areas are crucial to measuring health outcome data and can be defined through a variety of methods (e.g., vector-based, network analysis, raster grid cells). Figure 4.1 illustrates hospital catchments in two scenarios. The first depicts non-overlapping catchments typically found in rural areas. The second depicts overlapping catchments found in dense urban areas or in jurisdictions such as the US where multiple hospitals compete for the same patient base.

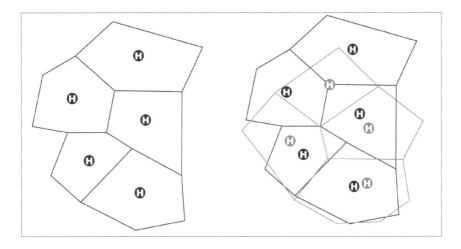

**Figure 4.1 Comparing hospitals with simple catchments to hospitals with overlapping catchments**

Appropriate hospital catchments are needed to assess the impact of health care restructuring and changes that specifically affect users' access on health outcomes. Specifically, health researchers need accurate denominator populations (e.g., total population counts) to link health outcomes with hospital services (Church et al. 1995). Ideally, rural hospital catchments would match the perceived geographic coverage and actual utilization patterns of hospitals.

Defining health service areas or catchments is problematic. Given Canada's uneven population distribution (i.e., the vast majority of Canadians live large metropolitan areas, with the remainder living in dispersed small regional centres or widely dispersed small rural and remote communities), many regions do not have populations large or dense enough to truly rationalize health service delivery without offloading the original costs on rural residents in terms of greatly increased travel-times (Church and Barker 1998).

The province of BC's geography makes it ideal for the creation of well-defined simple catchments for rural hospital services. BC is notoriously mountainous and has varying topography throughout. As a result *as the crow flies* catchments based on distance alone are inappropriate. In BC, different aerial (crow-fly) distances are used to provide minimum requirements of accessibility for acute care health services; however, these may create unrealistic catchment assumptions given the travel time obstacles presented by topographic features such as mountains.

The distance calculation methodology outlined in a 2002 BC Ministry of Health Services and Health Planning document equated 50 km of *aerial* distance with one hour of surface travel time. The population of rural BC was organized within catchment polygons of 50 km for emergency services, 100 km for acute inpatient services and 250 km for specialty services (Planning 2002). This approach to catchment definition does not account for elevation, speed limits, or even distance from roads, and therefore fails to describe with any veracity the likelihood that an individual will use a hospital or be able to receive care in a timely fashion. Such use of fixed vector polygons to describe catchment areas is an example of the *tyranny of zonation* in which fixed boundaries artificially depict a binary relationship between access and non-accessibility (Martin et al. 2002).

In many regions, utilization data is used to create catchments. In other words, administrators examine *which* people from *which* regions seek services from *which* hospitals. These usage patterns determine the catchments. Whist this method can be appropriate for dense urban populations, it works poorly for dispersed rural populations, especially when the terrain does not facilitate travel equally in all directions. In this chapter, urban population concentrations in the heavily populated south-western corner of the province of BC are deliberately ignored as multiple points of entry to the health care system exist. As a result, travel time is not a significant factor in determining access to health care. Instead the focus is on rural BC in order to demonstrate the utility of geographically defined catchments for health care access and the differential effect of service closures on different communities.

## GIS and Calculation of Catchments Based on Travel Time Distance

In rural BC *de facto* health service catchment areas have emerged because there is normally only one natural choice for patients to make, and it is dependent on travel time and proximity. The commonly used population-to-physician ratio is a primary care health services measurement tool that requires determination of a catchment population; however, the catchment areas used change with the researcher responsible for catchment definition (Pong and Pitblado 2001). Moreover, the resultant catchments are frequently unrelated to the circular 50, 100 or 250 km zones suggested by the BC Ministry of Health Services and Planning.

Catchment definition using GIS is frequently performed by use of vector as opposed to raster data structures. Vector refers to the use of polygons to define areas while raster is based on gridded cells that exhaustively describe the landscape. The most common method is to use Voronoi or Theissen polygons to tessellate (or subdivide into cells) the landscape.

A number of methods have been developed to calculate hospital catchments using vector analysis. One recent study used isochrones or equivalent drive times to identify non-urban populations. Catchments for health services were then calculated based on travel times to hospitals (Ellehoj et al. 2006). Least-cost analysis – a form of vector based catchments – was recently used to illustrate that hospital rationalization in New Zealand resulted in longer driving times for rural residents (Brabyn and Beere 2006). Small area analysis (SAA) can be also be used to create service zones. In another study SAA and GIS were used to visualize catchments and compare density of urban and rural zones of care (Klauss et al. 2005). David Martin, a UK researcher with considerable expertise in this area (Martin et al. 2002; Martin and Atkinson 2001; Martin and Williams 1992) created a population surface to integrate public and private transport to health care facilities to determine access times. In this case, the authors wanted to address the problem that access times are usually based on private car travel rather than public transit (Martin et al. 2002). In North America, where public transit rarely extends to rural areas, this innovation is not as relevant but nevertheless speaks to the imperative that understanding access is key to understanding: (1) utilization and (2) health outcomes.

Network analysis, one vector-based method of definition, is based on travel times, usually by road. Schuurman et al. (2006) developed a methodology using GIS to model hospital catchments in BC's rural and remote areas. The methodology uses Network Analyst, an extension to ESRI's ArcGIS software environment to model catchments around hospitals. Data were obtained from a travel-time road network for the entire province that permitted accurate distance modeling. The development and use of this methodology is based on the premise that understanding patient travel patterns is key to responsible location allocation policy for health care services. Driving time to care is the most accurate means of determining access in rural areas (Jordan et al. 2006).

## Using Catchments to Model Rural Access to Primary Health Care Services in BC

Catchments based on travel time reflect actual time on the ground – or in the car – required to reach a hospital or health service. As mentioned above, the province of BC has based its accessibility standards on time to definitive care but uses circular as the crow flies catchments to calculate travel time. The advantage of the Schuurman et al. (2006) catchment methodology is that it enables the development of catchments based on travel time as calculated from an accurate road network database. Figure 4.2 depicts circular catchments and travel time catchments for the eastern Kootenay region of BC. The Kootenays are mountainous regions with long valleys that run north-south along low elevation points between mountains. The roads are naturally contiguous with the valleys. Figure 4.2 illustrates that circular catchments are a poor alternative to travel time catchments. The latter can precisely determine where the cut-off point for access to a particular hospital lies, thereby allowing policy makers to determine the number of people affected by poor or no access and to identify poorly served areas of a health delivery service area. In this case, the population outside of one hour catchments is higher when travel time catchments rather than circular buffers are used. Note also that the catchments based on travel time are shaped like octopi, following the road network and the physical geography of the mountain valleys.

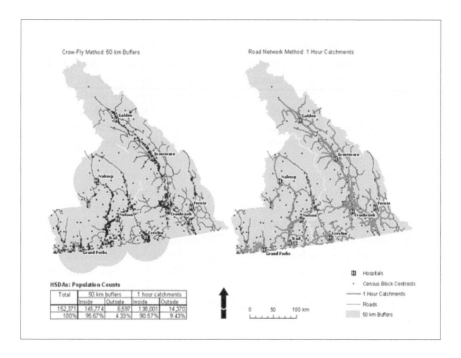

**Figure 4.2  East Kootenay and Kootenay/boundary health service delivery areas (HSDAs): access to hospitals with 24 hour emergency services**

*Accessing Rural Maternity Care*

Rational catchments optimize service planning for specific health services. Rural maternity care, for example, is a time sensitive service and adverse outcomes are frequently exacerbated by distance to care. Planning for rural areas, however, is thwarted by the high costs of maintaining obstetrics services within one hour driving time for a highly disparate population (Kornelsen and Grzybowski 2005a, 2005b). Problems of distance and cost are further exacerbated by a lack of central planning that examines this issue in particular.

In the Canadian province of BC, there are 72 hospitals outside the major cities of Vancouver and Victoria. Many of these hospitals provide only limited services.

As a result, patients are frequently referred to more distant tertiary centres. Only eight hospitals outside the large metropolizes, for instance, offer partial trauma services and only two offer full trauma services. Likewise, 57 hospitals offer limited maternity services while 23 have full obstetrics – at least part of the time. The number of rural residents in BC who live within one hour of any hospital is quite high as reflected in Figure 4.3. Almost 93 percent of a very dispersed population falls within one hour of some hospital service. This includes, however, hospitals without even an Emergency Department or Intensive Care Unit (ICU). By comparison, when the catchments for rural hospitals with full obstetric service are calculated, the number of people within one hour drops precipitously, to the extent that the vast majority of rural residents are outside of one hour to specific services (e.g., cardiac rehabilitation or maternity services).

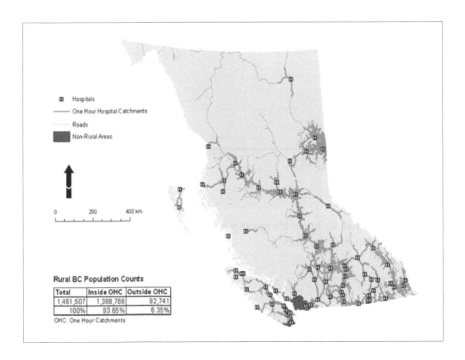

**Figure 4.3  British Columbia's rural population within one hour of any hospital services**

Figure 4.4 illustrates the extent of isolation from obstetric services for rural women. This follows a trend of hospital closure in Canada over the past two decades (Rourke 1998). There have been 17 closures of maternity services in BC since 2000 alone (Klein et al. 2002). In 2007, almost 24 percent of rural women were outside of one hour to obstetrical service. This has implications for families planning to have children and for those who are faced with possible high-risk pregnancies. It places the burden of responsibility for care on the patient. Many patients voluntarily re-locate during the final trimester to be closer to an obstetrician – at their own expense (Kornelson and Grzybowski 2005).

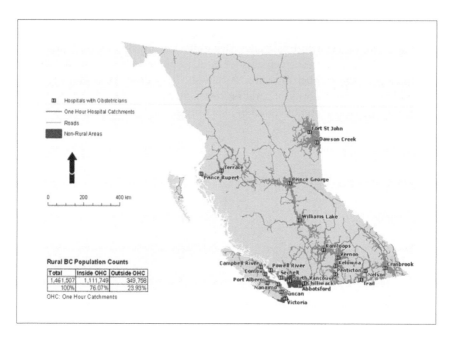

**Figure 4.4  British Columbia's rural population within one hour of obstetric services**

*Accessing Rural Trauma Care*

Trauma services are distributed across 14 hospitals in the province of BC. Of those, four are in the metropolitan south (e.g., the cities of Vancouver and Victoria). It is axiomatic in trauma research that treatment within one hour of traumatic injury decreases adverse outcomes (Branas et al. 2000; Kivell and Mason 1999; Pierce et al. 1999). Though the one-hour rule has exceptions, the principle that rapid access to care increases survival remains. A histogram of mortality associated with trauma normally has three prominent bars. The first peak represents those deaths that occur at the time of injury. Those injuries could only be mitigated by prevention rather than better care. The second significant peak occurs at about six hours after the injury. Rapid access to definitive care has a significant influence on this population, especially in avoiding massive loss of bodily fluids. Here time to definitive care has the greatest influence. The third distinct bar in the morality histogram associated with trauma occurs at about three weeks after the injury. Again rapid access to care influences this population as well. Clearly time to care is a significant factor in reducing the burden of injury on rural and remote populations.

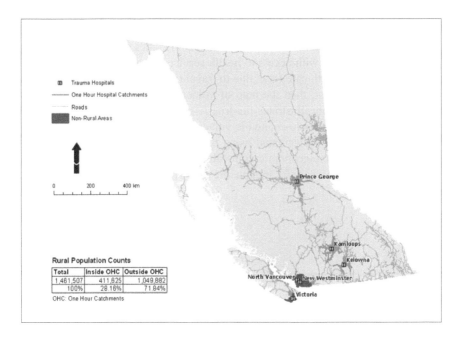

**Figure 4.5 British Columbia's rural population within one hour of trauma services**

In BC, there are two tertiary trauma centres outside the metropolitan areas of Vancouver and Victoria and eight district trauma facilities. District trauma facilities typically include an ICU, orthopaedics and an operating theatre. Full tertiary services will include neurosurgery. The latter are located in Kelowna and Kamploops. District trauma facilities are distributed throughout the province but primarily in the south, the exception being Prince George. An analysis of the distance of rural populations to definitive trauma care is illustrated in Figure 4.5. It is clear that a large proportion of the population – almost 72 percent – is farther than one hour to care. This has significant implications for morbidity and mortality in traumatic injury.

## Social Vulnerability Compounds Isolation

Rural residents are known to experience poorer health outcomes than their urban counterparts (Baird and Wright 2006). Certainly isolation is an acknowledged factor in this equation, but social vulnerability also plays a role (Braddock et al. 1994). Outcomes for injury are known to be worse among those in the lowest socioeconomic quintiles (Pappas et al. 1993; Wong et al. 2002; Cubbin et al. 2000). Poor SES likely compounds the risk of negative outcomes in rural areas. This is partly due to problems with access to health services in general, including those providing primary health care (e.g., due to travel time and lower car ownership). If SES is accounted for, then the map of vulnerability amongst rural residents is altered and ultimately illustrates the true risk associated with rurality. Figure 4.6 illustrates a map of socio-economic vulnerability in BC – linked to access to definitive trauma care – calculated using the VANDIX deprivation index (Bell et al. 2007). When socio-economic vulnerability is linked to rurality, specifically distance from definitive care for trauma, the effect is profound.

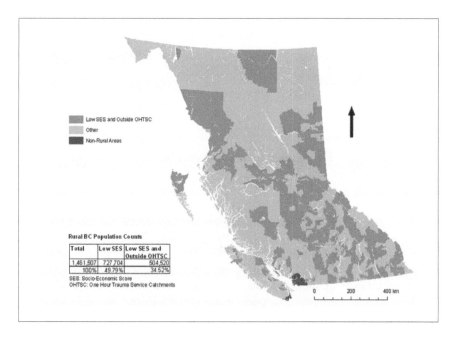

**Figure 4.6  Areas with low SES and outside one hour trauma service catchments in rural British Columbia**

Likewise, figure 4.7 illustrates rural regions that are more than one hour from rural maternity services and in the lowest two socio-economic quintiles. Again, northern populations in the province are more profoundly affected.

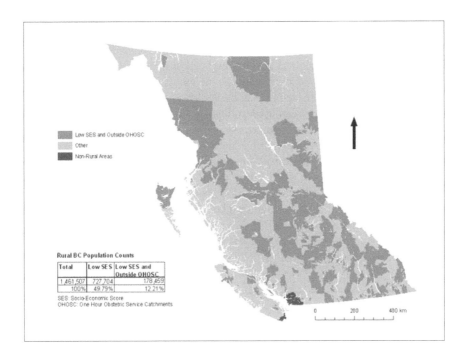

**Figure 4.7  Areas with low SES and outside one hour obstetric service catchments in rural British Columbia**

## Concluding Discussion

Physical distance to primary health care and related services is associated with real health risks for many rural Canadians. Certainly this is also true for other countries with dispersed rural populations (Brabyn and Skelly 2002; Brabyn and Beere 2006). While rurality is a risk factor, distance to care can be mitigated by careful health planning that uses real travel time catchments to make allocation decisions with the goal of minimizing distance for the largest numbers of people. Moreover, the role of socio-economic vulnerability and related risk must be recognized by policy makers, especially in rural settings where distance compounds social vulnerability.

The figures above have illustrated: (1) that rural health catchments for hospitals should be rationally defined, specifically based on travel time; (2) that rural residents have higher levels of increased distance to care; and (3) that socio-economic vulnerability can exacerbate risk associated with distance from care. Each of these factors can be integrated into primary care health services rationalization programs with the goal of enhancing health outcomes amongst rural residents.

Primary health care is based on equitable access to a range of health services, including maternity and trauma care. Clear lack of equitable access to these and other primary health care services (particularly those not delivered in local primary care clinics) contributes to rural vulnerability and more adverse health outcomes. These effects are compounded by other factors associated with rurality including lower SES as well as other factors (e.g. higher smoking rates).

In this chapter GIS has been used to illustrate the effects of distance on equity of access to specific essential components of primary health care using the case of rural BC. In the province of BC, there is clear and unequivocal evidence that rural and remote residents do not enjoy equitable access to these primary health care services. These GIS results constitute evidence for health planners involved in restructuring efforts. They are also a potentially powerful means to communicate the implications of reduced hospital facilities on rural residents and even on primary care services in that these service providers may ultimately bare the burden for lessened access to trauma and maternity care and other forms of specialist 'first contact' services.

## References

Adams, E.K. and Wright, G.E. (1991), 'Hospital Choice of Medicare Beneficiaries in a Rural Market: Why Not the Closest? ', *Journal of Rural Health* 7, 134–52.
Arcury, T.A., Gesler, W.M. and Preisser, J.S. et al. (2005a), 'The Effects of Geography and Spatial Behavior on Health Care Utilization Among the Residents of a Rural Region', *Health Services Research* 40, 135–55.
Arcury, T.A., Preisser, J.S., Gesler, W.M. and Powers, J.M. (2005b), 'Access to Transportation and Health Care Utilization in a Rural Region', *The Journal of Rural Health* 21, 31–38.
Baird, A.G. and Wright, N. (2006), 'Poor Access to Care: Rural Health Deprivation?', *British Journal of General Practice* 56, 567–68.
Barer, M.L., Morgan, S.G. and Evans, R.G. (2003), 'Strangulation or Rationalization? Costs and Access in Canadian Hospitals', *Longwoods Review* 1, 10–19.
Bell, N., Schuurman, N. and Hayes, M.V. (2007), 'Using GIS-based Methods of Multicriteria Analysis to Construct Socio-economic Deprivation Indices', *International Journal of Health Geographics* 6.
Brabyn, L. and Beere, P. (2006), 'Population Access to Hospital Emergency Departments and the Impacts of Health Reform in New Zealand', *Health Informatics Journal* 12, 227–37.

Brabyn, L. and Skelly, C. (2002), 'Modeling Population Access to New Zealand Public Hospitals', *International Journal of Health Geographics* 1, 1–10.

Braddock, M., Lapidus, G. and Cromley, E. et al. (1994), 'Using a Geographic Information System to Understand Child Pedestrian Injury', *American Journal of Public Health* 84, 1158–61.

Branas, C.C., Mackenzie, E.J. and Revelle, C.S. (2000), 'A Trauma Resource Allocation Model for Ambulances and Hospitals', *Health Services Research* 35, 489–507.

British Columbia Ministry of Health Services and Health Planning (2002), *Standards of Accessibility and Guidelines for Provision of Sustainable Acute Care Services by Health Authorities* (Victoria: Province of British Columbia).

Campos-Outcalt, D., Bay, C., Dellapena, A. and Cota, M.K. (2003), 'Motor Vehicle Crash Fatalities by Race/Ethnicity in Arizona, 1990–96', *Inj Prev* 9, 251–56.

Church, J. and Barker, P. (1998), 'Regionalization of Health Services in Canada: A Critical Perspective', *International Journal of Health Services* 28, 467–86.

Church, J., Saunders, L.D., Wanke, M.I. and Ong, R. (1995), *Organizational Models in Community-based Health Care: A Review of the Literature* (Ottawa: Health Canada).

Cromley, E.K. and McLafferty, S.L. (2002), *GIS and Public Health* (New York: Guilford Press).

Cubbin, C., Leclere, F.B. and Smith, G.S. (2000), 'Socioeconomic Status and the Occurrence of Fatal and Non-fatal Injury in the United States', *American Journal of Public Health* 90, 70–77.

Ellehoj, E., Tepper, J., Barrett, B. and Iglesias, S. (2006), 'Research Methodology for the Investigation of Rural Surgical Services', *Canadian Journal of Rural Medicine* 11, 187–94.

Faelker, T., Pickett, W. and Brison, R.J. (2006), 'Socioeconomic Differences in Childhood Injury: A Population Based Epidemiologic Study in Ontario, Canada', *Injury Prevention* 6.

Furuseth, O. (1998), 'Service Provision and Social Deprivation', in B. Ilbery (ed.), *The Geography of Rural Change* (Longman: Harlow).

Gilthorpe, M.S., Wilson, R.C. and Bedi, R. (1997), 'A Sociodemographic Analysis of Inpatient Oral Surgery: 1989–1994', *British Journal of Oral and Maxillofacial Surgery* 35, 323–27.

Guagliardo, M.F. (2004), 'Spatial Accessibility of Primary Care: Concepts, Methods and Challenges', *International Journal of Health Geographics* 3, 1–16.

Hartley, D. (2004), 'Rural Health Disparities, Population Health, and Rural Culture', *Rural Health and Health Care Disparities* 94, 1675–78.

Hotson, K.E., MacDonald, S.M. and Martin, B.D. (2004), 'Understanding Death and Dying in Select First Nations Communities in Northern Manitoba: Issues of Culture and Remote Service Delivery in Palliative Care', *International Journal of Circumpolar Health* 63, 25–68.

Jordan, H., Roderick, P., Martin, D. and Barnett, S. (2006), 'Distance, Rurality and the Need for Care: Access to Health Services in South West England', *International Journal of Health Geographics* 3, 1–9.

Joseph, A.E. and Bantock, P.R. (1982), 'Measuring Potential Physical Accessibility to General Practitioners in Rural Areas: A Method and Case Study', *Social Science & Medicine* 16, 85–90.

Kivell, P. and Mason, K. (1999), 'Trauma Systems and Major Injury Centres for the 21st Century: An Option', *Health & Place* 5, 99–110.

Klauss, G., Staub, L., Widmer, M. and Busato, A. (2005), 'Hospital Service Areas – A New Tool for Health Care Planning in Switzerland', *BMC Health Services Research* 5, 33.

Klein, M., Johnston, S., Christilaw, J. and Carty, E. (2002), 'Mothers, Babies, and Communities: Centralizing Maternity Care Exposes Mothers and Babies to Complications and Endangers Community Sustainability', *Canadian Family Physician* 48, 1177–9, 1183–5.

Kornelsen, J. and Grzybowski, S. (2005a), 'Is Local Maternity Care an Optional Service in Rural Communities?', *J Obstet Gynaecol Can* 27, 327–29.

Kornelsen, J. and Grzybowski, S. (2005b), *Rural Women's Experiences of Maternity Care: Implications for Policy and Practice* (Ottawa: Status of Women Canada).

Larson, E.H., Hart, L.G. and Rosenblatt, R.A. (1997), 'Is Non-Metropolitan Residence a Risk Factor for Poor Birth Outcome in the US?', *Social Science & Medicine* 45, 171–88.

Liu, L., Hader, J. and Brossart, B. et al. (2001), 'Impact of Rural Hospital Closures in Saskatchewan, Canada', *Social Science & Medicine* 52, 1793–1804.

Martin, D. and Atkinson, P. (2001), 'Investigating the Spatial Linkage of Primary School Performance and Catchment Characteristics', *Geographical and Environmental Modelling* 5, 67–83.

Martin, D. and Williams, H.C.W.L. (1992), 'Market-area Analysis and Accessibility to Primary Health-care Centres', *Environment and Planning A* 24, 1009–19.

Martin, D., Wrigley, H., Barnett, S. and Roderick, P. (2002), 'Increasing the Sophistication of Access Measurement in a Rural Health Care Study', *Health & Place* 8, 3–13.

Meade, M.S. and Earickson, R.J. (2000), *Medical Geography* (New York: Guilford Press).

Miles-Doan, R. and Kelly, S. (1995), 'Inequities in Health Care and Survival after Injury Among Pedestrians: Explaining the Urban/Rural Differential Increasing the Sophistication of Access Measurement in a Rural Health Care Study, *J Rural Health* 11, 177–84.

Orpana, H.M. and Lemyre, L. (2004), 'Explaining the Social Gradient in Health in Canada: Using the National Population Health Survey to Examine the Role of Stressors Increasing the Sophistication of Access Measurement in a Rural Health Care Study, *Int J Behav Med* 11, 143–51.

Pappas, G., Queen, S., Hadden, W. and Fisher, G. (1993), 'The Increasing Disparity in Mortality Between Socioeconomic Groups in the United States, 1960 and 1986 Increasing the Sophistication of Access Measurement in a Rural Health Care Study, *The New England Journal of Medicine* 329, 103–09.

Pereira, G.J. (2005), 'Palliative Care in the Hinterlands: A Description of Existing Services and Doctors' Attitudes', *Australian Journal of Rural Health* 13, 343–47.

Pierce, J.A., Ray, L.U., Wood, S. and Murrin, P. (1999), 'Creating and Using a Geographic Information System for Motor Vehicle-related Injury Surveillance in San Diego County', *Topics in Emergency Medicine* 21, 26–31.

Pong, R. and Pitblado, J.R. (2001), 'Don't Take 'Geography' for Granted! Some Methodological Issues in Measuring Geographic Distribution of Physicians', *Canadian Journal of Rural Medicine* 6, 103–112.

Romanow, R.J. (2002), *Building on Values: The Future of Health Care in Canada. Final Report of the Commission on the Future of Health Care in Canada* (Ottawa: National Library of Canada).

Rosenberg, M.W. (1988), 'Linking the Geographical, the Medical and the Political in Analysing Health Care Delivery Systems', *Social Science and Medicine* 26, 179–86.

Rourke, J.T.B. (1998), 'Trends in Small Hospital Obstetrical Services in Ontario', *Canadian Family Physician* 44, 2117–24.

Ryan-Nicholls, K.D. (2004) 'Health and Sustainability of Rural Communities', *Rural and Remote Health* 242.

Schuurman, N., Fiedler, R.S., Grzybowski, S.C. and Grund, D. (2006), 'Defining Rational Hospital Catchments for Non-urban Areas Based on Travel-time Health and Sustainability of Rural Communities,' *International Journal of Health Geographics* 5:43.

Sin, D.D., Svenson, L.W. and Man, S.F. (2001), 'Do Area-based Markers of Poverty Accurately Measure Personal Poverty?' *Can J Public Health* 92, 184–87.

Sutherns, R., McPhedran, M. and Haworth-Brockman, M. (2003), *Rural, Remote and Northern Women's Health: Policy and Research Directions. Final Summary Report* (Canada: Centres of Excellence for Women's Health).

Swindlehurst, H.F., Deaville, J.A., Wynn-Jones, J. and Mitchinson, K.M. (2005), 'Rural Proofing for Health: A Commentary', *Rural Remote Health* 5, 411.

Veenstra, G., Luginaah, I., Wakefield, S., Birch, S., Eyles, J. and Elliott, S. (2005), 'Who You Know, Where You Live: Social Capital, Neighbourhood and Health', *Social Science & Medicine* 60, 2799–2818.

Wardman, D., Clements, K. and Quantz, D. (2005), 'Access and Utilization of Health Services by British Columbia's Rural Aboriginal Population', *Int J Health Care Qual Assur Inc Leadersh Health Serv* 18, xxvi-xxxi.

Watson, D.E., Krueger, H., Mooney, D. and Black, C. (2005), *Planning for Renewal – Mapping Primary Health Care in British Columbia, Vancouver* (British Columbia: Centre for Health Services and Policy Research).

Wong, M.D., Shapiro, M.F., Boscardin, W.J. and Ettner, S.L. (2002), 'Contribution of Major Diseases to Disparities in Mortality', *The New England Journal of Medicine* 347, 1585–92.

Chapter 5

# The Role of Scale in Conceptualizing Primary Health Care Practice: Considering Social and Institutional Structures and Systems

Janine L. Wiles and Mark W. Rosenberg

Three trends have dominated the geographical organization of primary health care (PHC) in recent decades. First, there has been the slow drift away from single practice locations to group practice locations. Secondly, in many jurisdictions (e.g., the United Kingdom, New Zealand and Canada) where the majority of the population receives PHC mainly paid for through some form of public health insurance, governments have explicitly used regional organization as a mechanism for planning and sometimes funding PHC. A third trend has been the challenge of replacing and attracting PHC workers (especially physicians) to locate in rural and remote locations.

   Each of these trends is a manifestation of the importance of geographic scale. It is recognized that by delivering PHC in group practices, economies of scale can be achieved thereby making the delivery of PHC more efficient. Implicit in this recognition, however, is also the notion of geographic scale – the geographic catchment area for patients will be larger for a group practice than for a single practice. Some of the key underlying questions regarding the regional delivery of PHC are the optimal number, size and geographic scale of regional health authorities. These questions lead governments to review, change and reconfigure health authorities over time. Implicit in trying to attract PHC workers to locate in rural and remote locations is the recognition that working in these locations means working over a much broader geographic region than one would normally work in an urban location. In other words, the geographic scale upon which PHC is delivered changes depending on where care is delivered.

   As geographers, we seek to understand how particular scales are shaped and transformed in response to tensions that exist between structural forces and the practices of human agents (Marston 2000). In this chapter, scale is understood as a key factor in the construction of geographies rather than simply as an outcome of social relations (Howitt 1998). For example, the structure of the system shapes practice and delivery of PHC, such as whether PHC systems are centralized or decentralized; the degree to which decision making and direction is policy-maker driven or provider led; and whether systems are framed around a population health approach or individual level issues, or around preventive PHC or PHC as first point of contact and gatekeepers to higher levels of care. Other issues include the tensions involved in public health approaches between broad structural change and individual behavioural change, and whether public and private forms of PHC operate at the same kinds of scale.

The remainder of the chapter is organized into three parts: (1) scale as theoretically conceived in human geography and social theory; (2) the importance of scale in understanding PHC; and (3) the implications for PHC services and delivery. For consistency and ease of understanding, we use the terms PHC and PHC workers except in directly quoting others who use other terms to denote services that are preventive and the first point of contact for illness and providers who deliver such care.

## What is Scale?

Debates around structure and agency have long been a part of inquiry in human geography, and scale itself is an outcome (and producer) of the tensions between structural forces and the practices of human agents (e.g., see Gregory 1981). Scale is one of the most foundational, but also most confusing, concepts in geography (Brenner 2001; Howitt 1998). Smith (2000) distinguishes between *cartographic scale*, or the level of abstraction on a map (where a small scale map shows a large area at the expense of detail, and a large scale map shows a small area in great detail); *methodological scale,* or the choice of scale made by a researcher gathering information, such as whether data is gathered at the level of the census tract, ward, or district; and *geographic scale*, or the dimensions of specific landscapes, such as the regional scale, the scale of a watershed, the urban scale, or the global scale. All of these scales are *conceptions* rather than inevitable, natural, or given – but the conceptualization of geographic scale follows specific processes in the physical and human landscape rather than abstractions (such as a census tract) lain over it (Smith 2000). In this chapter, it is primarily geographical scale in which we are interested.

Howitt (1998; 2002) suggests that geographic scale comprises elements of size, level, and relations. Each of these three is helpful but inadequate on its own to explain the concept of scale. *Size* is useful in the sense of spatial extent, or a unit of measurement, such as the area of a health catchment, as described above. But ultimately size is too reductive as scale also contains spatial, temporal and social aspects. The idea of scale as a *level* (often described using the example of Russian dolls, one inside the other) is helpful for thinking of the nested idea of scale or increasingly wider levels each encompassing a greater amount of complexity. Brenner, for example, advocates for thinking about the relationship between different spatial units within a 'multi-tiered, hierarchically configured, geographical scaffolding' (Brenner 2002, 600). Smith (2000, 725) also suggests we recognize a 'loose hierarchy' of scales as a way to provide a conceptual language for analysing scale, to express the fact that the production of scale is 'not arbitrary or voluntaristic but is to a considerable degree ordered.' This idea is useful but too often does not emphasize the complexity of scale and worse, leads to the assumption of scale as a kind of vertical hierarchy.

Howitt (1998; 2002) advocates for thinking about scale as *relation*, in that scale is better understood dialectically rather than hierarchically. That is, we need to be able to understand the relationships *among* complex systems (such as health services, social and cultural values, and political systems) in order to make sense of a particular scale. While many elements may remain consistent in a geographic analysis that spans different geographical scales, what changes is 'the relationships we perceive between them and the ways we might emphasize specific elements for analytic attention' (Howitt 1998, 55). For example, we need to recognize that boundaries are set around a scale such as 'the local' for analytic purposes only, recognizing that 'external' linkages are actually 'internal' relations that also comprise part of the local (Howitt 2002). By thinking of scale as all three (size, level, and relations) we get a sense of larger scale entities that contain smaller scale entities, but that the larger scale entities are at the same time part of the smaller entities (Howitt 2002). They are mutually embedded. That is, the regional scales contain aspects of bodily and local scales just as they contain aspects of global and other scales.

Scale is better thought of as a metaphorical label or even a process, rather than as a discrete 'thing' that has power in its own right (Howitt 2002; see also Brenner 2001). Scale is a way of framing reality, rather than an external fact. It is also important that scale not be conflated with other key geographical concepts such as space, place, or time (e.g., place is often thought of as a particular location like a household or a community, but when we think of the household or local scale we need to do so in the sense of the relationship [and mutual embeddedness] of this scale with other scales). Geographic scale is different from scale in the sense of intensity, although this idea is informative in that there is an inherent sense of relatedness about scale. A point on a scale of intensity or gradation such as a Likert scale would have no meaning without a sense of how it relates to other points in the scale.

A number of researchers and activists have noted the political implications of scale. As Smith (1992) has argued, it is scale which 'defines the boundaries and bounds the identities around which control is exerted and contested.' 'Scale jumping,' for example, is the process where political claims and power established at one geographical scale are expanded to another (Smith 2000). In another sense, understanding the role of scale in political action is important, for example in thinking about how to link social movements or imagine and highlight social alternatives (Marston et al. 2005). While there is 'progressive potential for action at lower levels of scales' such as the household or the community, activism at such small scales can also result in barriers to taking efforts to larger scales (Staeheli 1994, 388). On the other hand, movements can also take advantage of crossing scales, by using the resources at one scale to overcome the constraints of another (Smith 2000). A community might be able to use opportunities at a national level when local structures do not work, for example.

It should be noted that several authors have argued that conceptualizing scale as a kind of vertical hierarchy is a misuse of the concept and should be avoided. There is a strong political argument against this kind of vertical hierarchalization which is seen as being disempowering, because the abstraction of phenomena or events to scales such as the 'global' tends to lead to an abstraction of the forces and agents of capitalism and globalization, and to a stabilizing of these abstractions so that they are understood to be unchangeable and factual. This has led to a tendency to focus on global scales and on the production of capital as if it were somehow abstract and disembodied, often at the expense of more 'everyday' processes of social reproduction and consumption, even though capital is an 'everyday' process (Marston 2000).

A range of alternatives to understanding scale as a vertical hierarchy have been suggested (Marston et al. 2005). Howitt (1998; 2002), for example, suggests a focus on the relational aspects of scale. Some have tried to invert the focus on vertical ideas of scale by approaching it from 'underneath', by focusing on the body or the 'local' as an entry point to examine broader social processes (Cox 1997). Others have argued for a re-emphasis on imagining horizontal networks of social processes that 'span space rather than cover it' as a way to overcome the rigidities in the way vertical understandings are applied (Leitner 2004). Marston and colleagues (2005) propose a 'flat' alternative to nested hierarchical views of scale which they reject along with horizontal 'out from here' versions. Masuda and Crooks (2007, 257) argue for an 'experiential approach' which privileges the role of human agency in supra-local encounters. They suggest that the local is not simply a piece of an aggregate global, but is a representative fraction of an unfathomably complex, yet lived-in world.

## Aspects of Primary Health Care Related to Scale

It can be argued that PHC and health care systems in general serve as one interface between individuals and the larger economic, political, and social systems in which they live. Health care is produced by system inputs (health professionals, medicines, facilities) that interact with the population through various processes (health consultations, surgeries, deliveries) and result in health outcomes (Macinko et al. 2003). PHC is unique because it tends to be a patient's point of entry into health care systems, and such care is at the centre of a network of providers working within and across primary, secondary, tertiary acute and social care boundaries. PHC delivers high quality, cost-effective care to patients (Schmittdiel et al. 2006), and plays an important role in controlling health expenditure (Bodenheimer 2006).

The *health needs* of a population are influenced by factors at both macro- and micro-levels. Macro-level health determinants include environments (e.g., the overall availability and quality of natural resources affected by national policies) and demographics (e.g., a greater number of deaths can be expected in countries with a higher percentage of older people or children, see Macinko et al. 2003). Other macro-level health determinants include socio-economic factors at the national level, such as Gross Domestic Product which is consistently found to be positively associated with better health outcomes (Macinko et al. 2003). Likewise more equitable distributions of wealth at a national, state, and even city level within countries are associated with better average health outcomes for individuals for a range of morbidity and mortality indicators (Dunn et al. 2005; Ross et al. 2005). At the micro-level, health determinants might be measured through per capita income, access to resources such as housing, education and nutrition, or factors such as alcohol and tobacco consumption. Micro-level factors influencing health might also be assessed through the accessibility of health care, as measured by average numbers of ambulatory care hospital visits, for example.

Accessibility is an important aspect of PHC, to facilitate receipt of care when it is first needed for all those who need it (Starfield 2001). For similar reasons, equity, or the principle that PHC should give priority to those most in need, and that PHC is seen as a means to redistribute wealth and access to resources in a population, is essential. Accessible, equitable PHC is one way to shift imbalances in health care, especially in regions/countries with massive inequalities and inequities. Several researchers argue that PHC is thought to be less costly towards individuals and more cost-effective to society, 'thus freeing up resources to attend to the health needs of the most disadvantaged' (Macinko et al. 2003, 832–3; Starfield 1994). This work shows that countries with strong PHC systems have more equitable health outcomes than those with systems oriented toward specialty care (see also Schmittdiel et al. 2006). Moreover, PHC is shown to be as effective as specialist care for health outcomes for some conditions (such as non-insulin dependent diabetes) and also less expensive (Greenfield et al. 1995). One study showed that for a range of 24 preventive and acute illness conditions, episodes-of-care that began with visits to an individual's PHC clinician, as opposed to other sources of care, were associated with reductions in expenditures (Forrest and Starfield 1996).[1] A more recent study (Shi et al. 2002) shows that good PHC experience (defined as enhanced accessibility and continuity) is associated with better self-reported general and mental health and reduces the adverse association of income inequality with general health (though not with mental health), especially in areas with highest income inequality.

*Structurally*, PHC is organized through a series of decisions, rules and policies embedded within the economic and political systems of the places in which

---

1   First contact with PHC physician was associated with reductions in expenditures of 53 percent overall, 60 percent for acute illnesses and 20 percent for preventive care. For 23 of 24 health problems studied, first contact care led to reductions in expenditures.

people live. The decision, rules, and policies are everything from what constitutes appropriate levels of training to qualify as a PHC provider, and the rules of the licensing body that provides the credentials for the delivery of PHC, to political decisions about whether providers will be paid through public health insurance or by the patient privately. Publicly- and privately-funded forms of PHC operate at quite different scales (individual and family, compared to community/region/ national level). For example, there is a difference between PHC initiatives aimed at individuals (usually around behavioural change) compared with those aimed at improving the health or access to health resources of groups (structural changes).

This difference is particularly evident when looking at the scale at which researchers in the US (where PHC is likely to be paid through private funds) think about primary care. For example, Safran et al.'s (2002) seven defining characteristics of PHC (which they measure on 11 summary scales), include access, continuity, integration, comprehensiveness, 'whole-person' orientation, clinical interaction, and sustained clinician-patient relationship. These appear to match a larger scale, group orientation to PHC practice. The characteristics are measured through surveys of individual patient satisfaction, however, suggesting a greater orientation towards individual perception of quality (e.g., patient perspectives on relationships with PHC physicians) (Safran et al. 2002). Sandy and Schroeder (2003) provocatively suggest that there is a growing distinction in US PHC between patient-centred care for wealthy consumers and family- and community-oriented bio-psychosocial models of care provided for the poor and working class, the latter being more closely related to 'family medicine models.' Indeed, compare the more clinician-oriented US approach with that of the Australian Primary Health Care Research Institute, a group of researchers who advocate defining PHC as (Sibthorpe et al. 2005, S52):

> ... socially appropriate, universally accessible, scientifically sound first level care provided by a suitably trained workforce supported by integrated referral systems and in a way that gives priority to those most in need, maximises community and individual self-reliance and participation and involves collaboration with other sectors. It includes health promotion, illness prevention, care of the sick, advocacy and community development.

Here, the emphasis is more on the structure and organization of the system, with a similar focus on equity in terms of priority to those with greatest need, an emphasis on participation in the deeper sense of involvement and partnership, and inter-sectoral collaboration.

Another related issue is whether service providers are paid salaries or fee-for-service. In a fee-for-service system, the impetus on professionals is to see as many patients as possible, and sometimes to conduct as many tests as possible. In contrast in a salary system, professionals may see fewer patients and place greater emphasis on overall health and well being. In a population-based funding system, there is likely to be greater emphasis on health promotion and illness prevention strategies. In a privately funded system and/or a system more oriented towards specialist acute services, there is a tendency for practitioners (including doctors and nurses) to work in more lucrative specialist services rather than generalist PHC (Bodenheimer 2006).

Funding arrangements thus also shape the spatial distribution of health providers and their practices because they have implications for recruitment and retention issues in rural and remote areas, in areas with high levels of socio-economic inequity and poverty, and in generalist PHC services.

Another issue inherently connected with scale is whether responsibility (e.g., for decision making, or funding) for PHC is centralized or devolved to regional or community levels. This will have implications for the ease of implementing change in knowledge and or technology for improving the health of populations, or developing general principles and targets for care provision. In a more centralized system, general approaches are easier to introduce. More centralized systems, however, may compromise local-level responsiveness and appropriateness, and the ability of local communities and groups to participate genuinely in decisions about the provision of PHC. Thus another factor influencing PHC is the degree to which decision making and direction is in fact policy-maker driven or provider-led (e.g., through managed funding), or informed by patient or 'consumer' participation (e.g., through election to health boards, participation in PHC organizations). Participation is frequently viewed as a key aspect of PHC. One set of community-oriented primary care researchers go so far as to argue that "primary care is the component of health services that addresses most of the health problems arising in a community, and when it is enhanced by a community orientation, it can be considered public health *at the local level*" (Gofin and Gofin 2005, 757, emphasis added).

Many researchers argue that there should be meaningful involvement of either (or both) individual patients and broader communities in the planning, implementation and maintenance of PHC (e.g., Sox 2003). Yet tension between a community-centred and a health-services-centred vision of PHC was a part of the World Health Organization's (WHO) development of the principles leading up to Alma-Ata (Litsios 2002) and still applies today. At the more health services end of the spectrum, the idea of participation may be focused on patient-centred care emphasizing the needs and preferences of individual patients especially in wealthy communities or societies (Showstack et al. 2003). For example, Donaldson et al. (1996, 5) define PHC as "the provision of integrated, accessible health care services by clinicians who are accountable for addressing a large majority of personal health needs, developing a sustained partnership with patients, and practicing in the context of family and community." This definition focuses on the provision of a particular kind of care (i.e., very generalized care) by clinicians with specific patients/consumers that happens to be in a community context (Schmittdiel et al. 2006). At the other end of the spectrum, participation, as an ideal characteristic of PHC, is extended to maximizing reliance on local resources (e.g., human resources, money, and materials), and appropriate technology adaptable to local needs and acceptable to local people. The latter is particularly found in more politicized areas, less wealthy communities and developing countries. Participation may mean full community control of planning, policy, management, and decision making power. For example, Starfield (2001, 454) defines PHC as:

> that aspect of a health services system that assures person focused care over time to a defined population, accessibility to facilitate receipt of care when it is first needed, comprehensiveness of care in the sense that only rare or unusual manifestations of ill health are referred elsewhere, and coordination of care such that all facets of care (wherever received) are integrated.

This definition focuses on the characteristics of a part of a health services system with respect to a defined population, rather than on the clinicians as providers of care. In general, her argument is that well-planned PHC promotes equity and the redistribution of resources.

Under specific decisions, rules and policies in any country, the consumption of PHC plays out at various geographic scales contingent upon the economic and social relations which mould peoples' everyday lives. At the local scale, where one accesses PHC will depend on where one works and the hours of work; where one goes to school and school hours or where one's children go to school and their school hours; whether one has access to a car, walks or uses public transit; one's sex and sexual orientation; one's marital status, family structure and household structure; one's income and education level; one's cultural background and the ability to communicate in the language of care providers; and whether one is an informal caregiver to others and the conditions of the care receiver. Out of this complex set of factors and relations that govern each individual's decision about where he or she will access PHC is created a geography of PHC based on where people live, work, go to school and shop. People find their PHC providers near their homes, where they work, where they or their children go to school or even in the shopping centres and precincts where they go to buy goods and services.

Overlaid on the local scale of primary care is a regional scale of *health organization*. Whether they are called regional health authorities, regional health networks, district health councils, local integrated health networks or something else, they are manifestations of higher levels of government attempting to create a 'rational' PHC system as part of a hierarchy of health care services (i.e., secondary, tertiary and even in some places quaternary health services mainly related to hospitals with increasingly complex levels of service and larger geographic catchment areas). The regionalization of health care places even more emphasis on structure as health regions are socially constructed and defined by rules, regulations and policies of the state with little regard for human agency (i.e., in regional health care systems, one is normally expected to seek their PHC and higher levels of care within the health region where one resides). Health regions are abstractions where only the most-informed are likely to know even the name of the region in which they live and even fewer are likely to have the knowledge to define the boundaries. Health regions are the products of the planners, policy-makers, and administrators where paradoxically the impacts on people's access to PHC can be significantly altered at any time and yet most of the population has no say in those rules, regulations or policies.

## Case Examples of Scale and Primary Health Care Systems

The previous section sketched out the critical theoretical issues defining PHC and how scale plays a role. In this section, we illustrate how many of the theoretical issues are played out in the reality of how PHC has been defined at the international, national, regional and local level in two case examples.

*International*

The WHO describes PHC as 'an integral part of the country's health system, of which it is the central function and main focus, and of the overall social and economic development of the community' (WHO 1978). WHO's Declaration of Alma-Ata (1978) outlined ten essential aspects of health and health care and the importance of primary care to achieving health for all (HFA). Embedded within the principles are five themes that have guided subsequent definitions of PHC and have also been the debating points for the differences in definitions. The five themes are: good health is a universal right; PHC can play a role in reducing inequalities and inequities within and between societies; PHC is the key to the creation of a productive workforce; individuals and governments have a role to play in the planning and delivery of PHC; and governments should work together at the international scale to provide PHC and HFA. In later discussions of the definitions of PHC, two of the themes (PHC as a key to a productive workforce and international cooperation) have generally been ignored. This reflects to a certain degree an emphasis on operational definitions.[2]

Principle VII[3] of the Declaration sets out in detail seven points that should define PHC. What is striking is that the seven points describe PHC as a comprehensive form of care. Later definitions mainly focus on much narrower, more selective factors (e.g., drawing a distinction between the role of PHC and public health). This comprehensive approach to PHC accounts for socio-economic and cultural determinants of health, identifies primary health needs, provides health care to the 'total community', and integrates preventive and curative care with significant community involvement (Gofin and Gofin 2005).

---

2   Note that the Declaration of Alma-Ata makes no reference to poverty. Litsios (2002) notes that this is because of diverging views amongst high level WHO staff about what role health care services could or could not play in alleviating poverty.

3   See Chapter 1 of this book for a verbatim listing of the principles of the Declaration.

Points 3 and 4 of Principle VII in the Declaration (that PHC must address some basic health needs and that responsibilities for PHC in a nation must be broader than the health sector) are now largely absent from discussions of PHC. Today in most developed countries, many of the basic needs are either seen as part of the public health care system (e.g., water and sanitation) or as part of other sectors of the economy (e.g., food security as an outcome of the agricultural sector) as distinct from PHC. In the bureaucratic realities of how developed countries organize their health care systems, PHC is, however, always seen as the responsibility of the ministry or department which has the mandate to oversee the health care system (although at least one early suggestion based on WHO research leading up to Alma-Ata was that PHC should be attached the Prime Minister's office or a similar post in each nation in order to allow for national level inter-sectoral co-operation). In fact, one might go so far as to argue that the division of government bureaucracies that arbitrarily allocate health, social development, education and the economy into various ministries and departments is one of the barriers to improving PHC and is exacerbated by the impacts of scale.

In addition to the interesting scalar implications contained within the Declaration itself, the history of the development of the Declaration and its subsequent manifestations are an excellent example of the nature of scale as size, level, and relation. Even PHC at this global scale is the outcome of a range of general and particular historical and contextual factors and a contingent process influenced by the personalities and politics of specific individuals and organizations. Throughout the 1950s and 1960s, the WHO had operated under what is often referred to as a 'vertical health approach.' This means a focus on technical solutions to specific diseases (such as malaria) delivered by highly trained health professionals and technicians (e.g., using Western-style hospital based health systems, vaccinations, and spraying insecticides). Such an approach was based on the assumption that the main diseases in poor countries were a natural reality that needed adequate technical solutions. As Cueto (2004) points out, this view contrasts with the recognition that diseases in less developed countries are socially and economically sustained, and need a political response. It was the latter recognition that informed the discussion that supported the development of the Declaration. This international agreement on a comprehensive and holistic, participatory, community-based approach to PHC is consistently referred to in the policy documents of many regions and countries and in a wide range of academic literature.

Litsios' (2002) personal reflection on the process leading up to the Declaration including the roles and values of individual personalities such as Mahler and Newell, and the political motivations and internal wrangling of organizations such as the Soviet Union or the United Nations (UN) is revealing. For example, as Director General of WHO for three consecutive five year terms, Mahler was a highly charismatic leader with strong links to a range of organizations, who held strong views on social justice, rational decision making based on research, and had a reputation for being 'radical' and 'anti-medical'. He argued for an ongoing process of change and innovation based on partnerships at 'the periphery' or from the grassroots supported by research, rather than led by a centralized headquarters, and felt that WHO should be working more at the country and regional levels. The Soviet Union in the context of the cold war, in contrast, was instrumental in pushing for a conference on health services development as a way to promote their own accomplishments in the field, perhaps as a way to demonstrate to less developed countries what their form of socialism could accomplish.

A wide range of other factors fed into both the existence and nature of the Declaration. For example, the holistic, integrated health associated with community development practiced as early as the 1940s in South Africa, which has subsequently been identified with the tradition of community-oriented PHC was also influential on the thinking of WHO officials such as Newell (Newell 1975; see also Gofin and Gofin, 2005; Mullan and Epstein 2002). The Canadian Lalonde report of 1974 emphasized social factors as key to determining health and signalled the beginnings of 'new' public health, social determinants of health, and later population health approaches (Canadian Department of National Health and Welfare 1974). Likewise, Cueto (2004) argues that the decolonization of many less developed countries and the growth of left-wing movements, as well as the growing popularity of grassroots, participatory approaches to development, played a role. For example, the Christian Medical Commission, which championed the training of village workers at the grassroots level, had strong relationships with WHO officers. Likewise, the global visibility of the barefoot doctor system (village health workers living in the communities they served, using preventive rather than curative services and a mix of traditional and Western medicine) was partly a function of the People's Republic of China joining the UN (and WHO), and the ability of this movement to capture the global imagination was hugely influential in the shift towards community-based initiatives (Cueto 2004).

Key players at the global level later adopted a more 'selective' definition of PHC with a few specific targets (such as growth monitoring or oral rehydration techniques), on the basis that the comprehensive Declaration was too idealistic and impossible to measure or assess. This was not a matter of abstract entities making decisions. The UN Children's Fund (UNICEF), for example, had been a major partner with the WHO in developing the comprehensive, holistic Alma-Ata Declaration, but a change in leadership led to a change in focus so that they later supported the more selective goal-driven model advocated by the Rockefeller Foundation and the World Bank (see Cueto 2004).

These examples illustrate the ways in which geographic scale involves relations between elements of complex and dynamic levels from the micro to the macro, from individuals through to federations of states, from long historical political processes through to a moment in a meeting. In a final note, we can observe that the Declaration warrants further consideration in terms of the way that scale was conceived in the statements. The document is inherently about scale. Explicitly, scale features in a hierarchical progression of relationships between PHC policies, programmes, and providers at the international, regional, national, and local levels. The local level is defined somewhat simplistically as 'community' which appears to be geographically oriented. But there is also an implicitly progressive focus on scale as relation, most notably in terms of the radical attempt to restructure relationships in the provision of health care (e.g., between systems focused narrowly and vertically on health services and structural reform towards inter-sectoral collaboration and holistic approaches to well-being).

*National, Regional and Local Scales: Canada*

In Canada, health care in general and PHC in particular is the constitutional responsibility of the provincial governments. This means that there is some variation province-by-province, but through a set of agreements since the 1950s, the federal government has sought to insert itself into the provision of health care and to reduce differences among the provinces. Without going into the details, the federal government initially used its spending power to encourage the provincial governments to create public hospital insurance and then medical insurance systems and ultimately public health insurance systems in 1966 through the Medical Care Act (see Naylor 1986; Torrance 1998 for details). These initiatives on the part of the federal government reached their zenith with the agreement between the federal and provincial governments in 1984 when the Canada Health Act was passed by the federal government (see Mhatre and Deber 1998).

The core of the Medical Care and Canada Health Acts (CHA) is that the federal government agreed to provide a substantial proportion of the funding of provincial health care and in return the provinces agreed that their health care systems would meet five key principles:

1. Universality – provincial health insurance systems would cover all eligible citizens in a province.
2. Comprehensiveness – provincial health insurance systems would cover all reasonable medical and hospital costs.
3. Accessibility – provincial health insurance systems would ensure that there are no economic barriers to health care and *de facto* the provincial governments would try to reduce geographic barriers to access wherever and as reasonably as possible.
4. Portability – people would be covered by their provincial health care plans anywhere in Canada.
5. Public Administration – public health insurance systems would be administered either by the provincial government directly or through a public agency that was accountable to the provincial government.

In essence, the Medical Care Act and CHA provided the broad parameters for the organization of health care in general and PHC specifically, in particular care at three scales. At the national scale, it meant that all eligible Canadians and legal immigrants had access to similar comprehensive care paid for through a public health insurance system regardless of where they lived in Canada. At the provincial scale, it meant that governments had considerable flexibility to organize health care in general and PHC in particular as they saw fit after 1966. In reality, provinces chose a range of strategies with respect to planning, funding, allocation, and regional organization of these functions. For example, the province of Alberta chose to create regional health authorities with wide-ranging autonomy to plan, fund, and allocate funds for services. In contrast in Ontario until 2006, although there were regional health authorities (i.e., district health councils [DHCs]) which had planning function, funding and allocation responsibilities, decisions for all intents and purposes were made at the Ministry of Health and Long-term Care; what some have called a 'command and control' system operated. At the local level in all provinces, physicians continued to practice under fee-for-service arrangements where the provincial government or public agency responsible for the provincial health insurance was the single payer. It was also the case that in 1984, physicians as PHC providers were to a large extent free to practice wherever they wanted, although some provincial governments created incentive programs to encourage physicians to practice in rural and remote locations specifically. Since 1984, changes have taken place at all three scales, so that today the delivery of PHC in Canada looks very different than it did in 1984.

What has led to the changes since 1984 are various economic, political, social, and medical forces operating at the three scales previously defined (see Burke and Stevenson 1998). Federally, in its efforts to reduce national deficits and debts, the government reduced the funds it transferred to the provinces over time until recently. With less federal funding as an overall proportion of all provincial funding on health care, provincial governments felt emboldened to reconfigure their health care systems in various ways. Most provincial governments were also struggling with deficits and debts. In addition, medical technology was becoming more complex, new pharmaceutical solutions were being developed, and the PHC workforce was aging. In sum, virtually every province faced and continues to face the same challenges: how to control health care spending; address a growing shortage of PHC workers especially in small town, rural and remote locations; and a general population which is simultaneously becoming older and more diversified ethnically.

At the provincial scale, all provinces have responded in a similar fashion although the timing of changes has differed as well as the details. All provinces have created regional health authorities of some form with varying levels of responsibility for planning health services and allocating funds to them. To address the issue of the aging of the PHC workforce especially in small towns and rural and remote areas, all provinces have programs that use incentives to attract primary care physicians to areas where there are shortages. Some provinces have tried disincentive programs to discourage physicians from practicing in areas where governments deem the supply adequate (mainly in large urban areas) and some provinces have tried to recruit physicians from other countries to practice in small towns and rural and remote areas. Some provinces have encouraged the development of community health centres (e.g., the province of Quebec reorganized PHC around *les centres local de services communautaires*), while others have encouraged group practices (e.g., the province of Ontario has a program, Family Health Teams, to encourage primary care physicians to organize into group practices). Yet another strategy employed by provincial government is to expand other parts of the PHC workforce. For example, the introduction of nurse practitioners and the re-introduction of midwifery in Ontario can be seen as responses to the lack of primary care physicians. Another trend has been to expand provincial health insurance coverage to services such as chiropractic, physiotherapy, and acupuncture. A final trend has been to encourage the elderly population to age-in-place and more specifically to remain in their homes as long as possible. To facilitate aging-in-place, a new class of primary care health worker has been developed (i.e., the home care worker). While medical care in the home is usually provided by a nurse, various non-medical services (e.g., homemaking and bathing) are being provided by home care workers.

While the above trends can be viewed positively as expanding the definition of PHC to reflect much of the theoretical discussion in the previous section, at the local scale the reality has often been quite different. At best, small towns, and rural and remote locations continue to struggle to recruit PHC workers or in some cases as the current PHC workforce retires, access to care is getting worse. Many of the innovations described in expanding the PHC workforce have had a much greater impact on urban areas where practitioners have tended to concentrate and thus have only exacerbated the inequalities in access to PHC between the larger urban areas and the small towns, rural and remote locations. Finally, because of the growing overall costs of health care resulting from the demand for new services, the costs of new medical technologies and provincial government expenditures on drug plans, providers of PHC often find that they do not have the resources to provide the services required to meet growing local demand. An example of this problem is that in Ontario, Community Care Access Centres (CCACs) are responsible for the delivery of home care, but because of the growth in the elderly population who require these services to age-in-place, many CCACs have had to reduce the non-medical services provided to ensure people get the medical services they require.

Ultimately, changes in the delivery of PHC in Canada since 1984 raise a number of more general issues related to scale. First, the lack of a uniform response by the provinces has resulted in greater differences in the delivery of PHC from province to province. Secondly, contrary to the CHA, which was intended to reduce inequality in access to PHC, greater inequality is developing among the provinces as each develops its own regional model and decides what services to fund and how best to deliver them. Thirdly, there is a 'knock-on' effect at the international scale where those provinces which are encouraging the relocation of PHC workers from developing countries to solve local shortages are exacerbating the problems of access to PHC in developing countries.

## Concluding Discussion

Scale is an issue which underpins, and is constructed through, the geographic organization and manifestation of PHC. Scale is a way of framing and organizing the world, which has tangible political consequences. Scale is a slippery concept, comprising not only a sense of *size* or *levels* of complexity, but also a way of understanding the *relationships* between complex systems. Common examples of scales frequently used by geographers and relevant to PHC include the body (or corporeal scale), the household, the local/community, urban, districts, national, regional, and the global. Yet at the same time, these scales must be understood as metaphorical concepts, developed and applied to help understand and organize the world around us. These scales are all inherently *related* to each other. What happens in PHC at a national scale, for example, is inherently connected to the behaviours and beliefs of individuals and to the provision of care at the local level whether by specific PHC providers or decision makers at the slightly larger district/regional scale, as well as by global, international events such as decisions and policies made through the WHO and other supra-national groups.

Yet we should not assume that scale is fixed or immutable, nor that scale is so abstract as to have no material consequences. The examples presented here illustrate these points very well. First, restructuring can dramatically alter the scale at which PHC is focused. For example, in Canada, the autonomy of provincial governments and the incredible range in geographic areas over which they have jurisdiction have led to variation in what constitutes PHC within a national framework of five principles that govern supposedly universal, comprehensive, accessible, portable and public health insurance systems.

Secondly, there are material and political manifestations of primary care at various scales, with very real consequences for PHC at other scales. The WHO Declaration of Alma-Ata proclaiming an holistic, comprehensive, contextually-sensitive and participatory approach to PHC, and subsequent practical and political struggles between global, regional, and local players over whether and how this could be achieved, have played out in different ways in regional (e.g., less wealthy compared to wealthier countries), national, and local contexts.

Thirdly, there are the unintended consequences of geographic scale. In seeking to improve accessibility to PHC in developed countries like Canada, the recruitment of PHC workers from developing countries has exacerbated the problems of accessibility to care in developing countries from where the practitioners originate.

Ironically, almost 30 years after the Alma-Ata Declaration, there is no consensus at any scale as to what PHC means. However, at virtually any geographic scale one might define, most would agree that without equitable access to PHC achieving HFA remains a global dream yet to be achieved.

## References

Arnstein, S. (1969), 'A Ladder of Citizen Participation', *American Institute of Planners* July, 216–24.

Bodenheimer, T. (2006), 'Primary Care – Will it Survive?', *New England Journal of Medicine* 335:9, 861–4.

Brenner, N. (2001), 'The Limits to Scale? Methodological Reflections on Scalar Structuration', *Progress in Human Geography* 25:4, 591–614.

Burke, M. and Stevenson, H.M. (1998), 'Fiscal Crisis and Restructuring in Medicare: The Politics of Health in Canada', in D. Coburn, C. D'Arcy, and G.M. Torrance (eds), *Health and Canadian Society. Sociological Perspectives*, 3rd edition (Toronto: University of Toronto Press), 597–618.

Canadian Department of National Health and Welfare (1974), *A New Perspective on the Health of Canadians/Nouvelle Perspective de la sante des Canadiens* (Ottawa, n.p).

Cox, K. (1997), *Spaces of Globalization: Reasserting the Power of the Local* (New York: Guilford Press).

Cueto, M. (2004), 'The Origins of Primary Health Care and Selective Primary Health Care', *American Journal of Public Health* 94:11, 1864–74.

Donaldson, M.S. and Yordy, K.D. et al. (eds) (1996), *Primary Care: America's Health in a New Era* (Washington DC: National Academy Press).

Dunn, J.R. and Burgess, W. et al. (2005), 'Income Distribution, Public Services Expenditures, and All-cause Mortality in U.S. States', *Journal of Epidemiology & Community Health* 59, 768–74.

Forrest, C.B. and Starfield, B. (1996), 'The Effect of First Contact Care with Primary Care Physicians on Ambulatory Health Care Expenditures', *Journal of Family Practice* 43:1, 30–3.

Gofin, J. and Gofin, R. (2005), 'Community-oriented Primary Care and Primary Health Care', *American Journal of Public Health* 95:5, 757.

Greenfield, S., Rogers, W. and Mangotich, M. et al. (1995), 'Outcomes of Patients with Hypertension and Non-insulin Dependent Diabetes Mellitus Treated by Different Systems and Specialties. Results from the Medical Outcomes Study', *Journal of the American Medical Association* 274:18, 1436–44.

Gregory, D. (1981), 'Human Agency and Human Geography', *Transactions of the Institute of British Geographers* 6:1, 1–18.

Howitt, R. (1998), 'Scale as Relation: Musical Metaphors of Geographical Scale', *Area* 30:1, 49–58.

Howitt, R. (2002), 'Scale and the Other: Levinas and Geography', *Geoforum* 33:3, 299–313.

Leitner, H. (2004), 'The Politics of Scale and Networks of Spatial Connectivity: Transnational Interurban Networks and the Rescaling of Political Governance in Europe', in E. Sheppard and R.B. McMaster (eds), *Scale and Geographic Inquiry* (Oxford: Blackwell), 236–55.

Litsios, S. (2002), 'The Long and Difficult Road to Alma-Ata: A Personal Reflection', *International Journal of Health Services* 32:4, 709–32.

Macinko, J. and Starfield, B. et al. (2003), 'The Contribution of Primary Care Systems to Health Outcomes within Organization for Economic Cooperation and Development (OECD) Countries, 1970–1998', *Health Services Research* 38:3, 831–65.

Marston, S. (2000), 'The Social Construction of Scale', *Progress in Human Geography* 24:2, 219–42.

Marston, S. and Jones III, J.P. et al. (2005), 'Human Geography without Scale', *Transactions of the Institute of British Geographers* 30:4, 416–32.

Masuda, J. and Crooks, V.A. (2007), '(Re)thinking the Scales of Lived Experience', *Area* 39:3, 257–58.

Mhatre, S.L. and Deber, R.B. (1998), 'From Equal Access to Health to Equitable Access to Health: A Review of Canadian Provincial Health Commissions and Reports', in D. Coburn, C. D'Arcy and G.M. Torrance (eds), *Health and Canadian Society. Sociological Perspectives*, 3rd edition (Toronto: University of Toronto Press), 459–84.

Mullan, F. and Epstein, L. (2002), 'Community-oriented Primary Care: New Relevance in a Changing World', *American Journal of Public Health* 92:11, 1748–55.

Naylor, C.D. (1986), *Private Practice, Public Payment: Canadian Medicine and the Politics of Health Insurance, 1911–1966* (Montreal-Kingston: McGill-Queen's University Press).

Newell, K.W. (ed.) (1975), *Health by the People* (Geneva: World Health Organization).

Ross, N., Dorling, D. and Dunn, J. et al. (2005), 'Metropolitan Scale Relationship Between Income Inequality and Mortality in Five Countries Using Comparable Data', *Journal of Urban Health* 82, 101–10.

Safran, D.G., Wilson, I.B. and Rogers, W.H. et al. (2002), 'Primary Care Quality in the Medicare Program: Comparing the Performance of Medicare Health Maintenance Organisations and Traditional Fee-for Service Medicare', *Archives of Internal Medicine* 162, 757–65.

Sandy, L.G. and Schroeder, S.A. (2003), 'Primary Care in a New Era: Disillusion and Dissolution?', *Annals of Internal Medicine* 138:3, 262–7.

Schmittdiel, J.A., Shortell, S.M. and Rundall, T.G. et al. (2006), 'Effect of Primary Health Care Orientation on Chronic Care Management', *Annals of Family Medicine* 4:2, 117–23.

Shi, L., Starfield, B. and Politzer, R. et al. (2002), 'Primary Care, Self-rated Health, and Reductions in Social Disparities in Health', *Health Services Research* 37:3, 529–50.

Showstack, J., Lurie, N. and Larson, E.B. et al. (2003), 'Primary Care: The Next Renaissance', *Annals of Internal Medicine* 138:3, 268–72.

Sibthorpe, B., Glasgow, N.J. and Wells, R.W. (2005), 'Questioning the Sustainability of Primary Health Care Innovation', *Medical Journal of Australia* 183:10, S52–3.

Smith, N. (1992), 'Contours of a Spatialized Politics: Homeless Vehicles and the Production of Geographical Space', *Social Text* 33, 54–81.

Smith, N. (2000), 'Scale', in R.J. Johnston, D. Gregory, G. Pratt and M. Watts (eds), *The Dictionary of Human Geography*, 4th edition (Oxford: Blackwell), 724–7.

Sox, H.C. (2003), 'The Future of Primary Care', *Annals of Internal Medicine* 138:3, 230–2.

Staeheli, L. (1994), 'Empowering Political Struggle: Spaces and Scales of Resistance', *Political Geography* 13:5, 387–91.

Starfield, B. (1994), 'Is Primary Care Essential?', *Lancet* 344, 1129–33.

Starfield, B. (2001), 'Basic Concepts in Population Health and Health Care', *Journal of Epidemiology and Community Health* 55, 452–4.

Torrance, G. (1998), 'Socio-historical Overview: The Development of the Canadian Health System,' in D. Coburn, C. D'Arcy, and G.M. Torrance (eds), *Health and Canadian Society. Sociological Perspectives*, 3rd edition (Toronto: University of Toronto Press), 3–22.

World Health Organization (1978), *The Alma-Ata Declaration* (Geneva: World Health Organization).

Chapter 6

# Cloaked Selective Primary Health Care? Local Observations of Rural Primary Health Care Clinics in Perú

Leah Gold

The work reported herein is part of a larger project examining clashes between biomedical and indigenous medical ideologies, and tensions and debates surrounding health care delivery in developing countries. In Perú, as in many developing countries, the biomedical practice of primary health care (PHC) is parachuted into areas where indigenous peasants (*campesinos*) have pre-existing, culturally embedded medical systems. My research includes narrative analysis from *campesinos* who regularly use indigenous medicine and their experiences within the local PHC system. In this chapter, I limit my scope to examine PHC through the establishment of the *Comités Locales de Administración de Salud*[1] (CLAS) system in Perú within the debate between comprehensive and selective PHC.

The debate between comprehensive and selective PHC, which reflects the 'tensions between social and economic approaches to population health and technology or disease-focused approaches' (Brown et al. 2006) is useful for this discussion.[2] While comprehensive PHC is characterized as low cost wide-ranging health care accessible to low-income and otherwise marginalized people, selective PHC is characterized by easily monitored vertical programs (sometimes temporary), a focus on disease and technology, and by efficiency trumping social solutions to ill health. The main goals of CLAS are characteristic of a comprehensive approach, and include the delivery of PHC as a poverty-reduction strategy, decentralizing health services, including community participation and fostering cultural sensitivity. This chapter evaluates two of the four of the CLAS goals, poverty reduction and community participation in health care delivery, in order to examine to what extent the CLAS system fulfills the goal of providing comprehensive PHC. I demonstrate that it more closely resembles selective PHC.

This research examines two rural CLAS clinics in villages near Cusco in the Sacred Valley – a popular tourist destination point, and gateway to the world

---

1   Local Health Administration Committees.

2   For an in-depth discussion of this debate see Cueto 2004, as well as articles in *Social Science & Medicine* 1988, 26:9.

famous archaeological site Machu Picchu. While Cusco itself is a city (albeit small) of cosmopolitan proportions, those who live in surrounding rural communities are mostly poor, subsistence-level *campesinos* who have access to a PHC clinic, but mostly rely on long-standing lay knowledge of medicinal plants passed down through generations. The following quotation came from an informant from the *communidad campesino* (rural community) of Añawi (depicted in Figures 6.1 and 6.2).[3] It gives a brief perspective of the history of health seeking and the presence of PHC in this community.

> Before, our grandparents cured themselves naturally with herbs, but when the Ministry of Health appeared, they obligated everyone to come to the clinic to be cured. If we didn't go, and someone in the community died, there would be a fine to pay, and if it wasn't paid, we weren't allowed to bury the deceased. So everyone lost this tool [of using natural herbs], and instead goes to clinics.

This quotation suggests that processes involving the negotiation of health care seeking occur within the politics of unequal power relations and economic disparities (Madge 1998). Añawayans were coerced through policies to visit the clinic in spite of pre-existing indigenous medicine in the region. Añawayans preferred their homes and community over the clinic as a place of healing. The evaluation of CLAS goals may offer insight into this preference. Here I argue that in its current model, the CLAS program cannot fulfil its original goals.

In this chapter, I briefly review health reform in Perú in order to contextualize the CLAS and provide an overview of my study in rural Perú. The remainder of the chapter examines CLAS initiatives of poverty reduction and community participation, which supposedly increase PHC coverage and empower local peoples. In particular, I consider how CLAS operationalizes the principle of poverty reduction and examine barriers to CLAS being able to broaden its coverage. I also examine how community participation affects the CLAS program, and whether or not it has led to increased community control of CLAS services.

---

3   This is a pseudonym.

**Figure 6.1  Añawi village and surrounding area**
*Source*: Author's photo.

**Figure 6.2  Typical houses in Añawi village**
*Source*: Author's photo.

## Health Reform in Perú and the CLAS Program

Due to inefficient services and high expenditures, health sector reform took place to varying degrees in different countries in Latin America after the economic crisis of the 1980s. Neoliberal economics were applied to cut social spending in a variety of sectors, including education, welfare and health services as a result of influence from international funders (Homedes and Ugalde 2005). The resulting cutbacks were particularly harsh on the low-income sectors (Asthana 1994). Privatization and decentralization were common to neoliberal reforms in Latin America, and, with the exception of Chile and Colombia, such reforms had limited success (Homedes and Ugalde 2005). Such reform in Perú took the form of structural adjustment programs that were designed to, among other things, cut inflation by cutting public expenditures, apply free market policies, reduce the economic role of the government and increase the role of the private sector.

Specifically in health services, selective strategies were favoured in order to reduce social spending, and increased economic efficiency trumped all the other objectives (Cortez 2000). Chile was the first Latin American country to reform their health care under the Pinochet government in the early 1980s (Puig-Janoy 2002) and most other countries followed suit in the 1980s and 1990s with great difficulty. Many countries have merely implemented reforms in a piecemeal fashion (Langer 2000; Homedes and Ugalde 2005). Perú's reforms were characterized as some of the harshest in Latin America (Ewig 2000) though they were typical of those undertaken by other Latin American countries and carried similar objectives of increased coverage, decreased inequality, decentralization and local participation (Iwami and Petchy 2002).

Neoliberal health reform in Perú began in 1991 under the Fujimori government through the introduction of user fees (Ewig 2000) when the health system, like the rest of the economy (Bowyer 2004), was in shambles. Fujimori was not interested in social policy, thus the health reform was more a response to the economic crisis than it was a desire for the improvement of the health system (Ewig 2002). Consequently reforms were focused on efficiency rather than equity (Iwami and Petchy 2002). The health care system in Perú has typically followed a segmented model where the various health care sectors (i.e., private, Ministry of Health [*Ministerio de Salud* or MINSA], and institutes for social security) are isolated and operate in a vertical manner (Londoño and Frenk 1997). The government has traditionally funded PHC for low-income sectors of the population, which mostly includes the rural indigenous and growing numbers of poor immigrating to suburban areas.

PHC in Perú is implemented through the MINSA's Basic Health for All Program (*Programa de Salud Básica Para Todos* – herein referred to as 'Basic Health') which was inspired by the World Health Organization's (WHO) *Health for all by 2000* approach declared at Alma-Ata in 1978, thus characteristic of comprehensive PHC. However, due to reports revealing that 70 percent of the poor were unhappy with the services offered through Basic Health (Cortez 2000) and that large sections of the population were still plagued with diseases of the poor (Altobelli and Pancorvo 2000), such as cholera, tuberculosis and diarrhoea, there was a recognized need to improve PHC. Financed through and built on the same comprehensive philosophies of Basic Health (Altobelli and Pancorvo 2000; Iwami and Petchey 2002), the CLAS program was born as a pilot project in 1994, and though inextricably linked, they differ (Ewig 2002). While Basic Health is 'centralized and targeted', CLAS is decentralized, thus 'targeting different political camps' (Ewig 2007, personal communication). Furthermore, there was resistance to CLAS by some regional authorities and local doctors as they saw community participation as a 'threat to their own power' (Ewig 2007, personal communication). Regardless, user satisfaction has been reported to be high (Cortez 2000) and the CLAS is considered to be a popular response to the restructuring of the health system (Iwami and Petchey 2002). To illustrate, in 1997, 12 percent of the health clinics were run by CLAS (Iwami and Petchey 2002) increasing to 25 percent in 2001 and 35 percent in 2004 (Future Generations 2004).

In its original mandate, the CLAS initiative had elements of comprehensive goals through its attempt to reduce poverty by servicing PHC needs of the poorest members of the population (Government of Perú 2000), exonerating service fees for the poor, broadening PHC coverage for the economically marginalized rural and urban populations, fostering cultural sensitivity, improving the *quality* of treatment, and prioritizing social programs (Cortez 2000). Furthermore, the CLAS initiative attempts to bestow more power onto local people through the process of decentralizing, which is a ministry wide policy (Iwami and Petchy 2002; MINSA 2004). As part of the decentralization process, CLAS clinics are promoted as legally private, not-for-profit entities (Altobelli 2002; Ewig 2002). However, they remain under state control through policy regulation by virtue of the fact that doctors and nurses are paid by the government, and that each clinic must apply for a three year contract with the government. Though clinics are described as not-for-profit, they rely on their profits to pay for materials and to increase the profile of the clinic by hiring specialists and providing health promotional programs (doctor's and health worker's wages do not increase). The interest for doctors and health workers in increasing profits is that if their clinic is successful, their contracts are renewed with the government and they receive positive evaluations from the community, thus keeping their jobs. Such positions are difficult to come by, and are thus competitive.

Clinics rely on the participation of elected community members for administration, making it appear like a grassroots initiative. It was thought that if communities gained more control of clinics, local peoples would be empowered, fiscal responsibility would improve (Altobelli and Pancorvo 2000; Cortez 2000; Noel 2005) and equality, coverage and the quality of services would increase (Altobelli and Pancorvo 2000; Cortez 2000). However, the CLAS is not without its critics. Through the little available literature on the CLAS system and my research, I evaluate its goals, and in particular examine how the CLAS system fits the paradigm of comprehensive PHC in the remainder of the chapter.

**Study Overview**

This research is situated within a health geographies framework, as described by Kearns and Moon (2002). I draw on critical cultural theories and ethnographic methods from critical medical anthropology. Ethnography is described as a constellation of qualitative methods with participant observation as its cornerstone (Herbert 2000), with the objective of providing insight into the 'everyday' as it occurs in natural settings (Kearns 2000). To that end, I employed a variety of ethnographic techniques. Specifically I employed semi-structured interviewing, participant observation and document analysis, drawing from numerous print and photographic sources.

From June 23 to September 24, 2004 I conducted a total of 53 interviews. These included one with a MINSA official in charge of *Programa de Administración de Acuerdos de Gestión*[4] (PAAG, or Administration Program for Management Agreement), seven with CLAS members,[5] 38 with residents of Añawi, and additional interviews held with local *curanderos*, herb vendors, and rural women from another community. The CLAS members were from two different clinics, which I refer to as Clinic A and Clinic B. Añawayans had access to Clinic B. All interviews were held in either Spanish or Quechua. Interviews conducted in Quechua with the aide of a translator were translated into Spanish on site.

The majority of my research focused on the health care seeking of Añawayans. I asked them about their perceptions and experiences with Clinic B to better understand their use of the clinic and possible barriers to its use. I also sought insight as to whether the CLAS was broadening its coverage of rural low-income sectors, and whether this contributed to a strategy of poverty reduction. Their answers offer insights into how the clinic may be better utilized and more accommodating for local peoples. To guide my interviews, I asked villagers what they do when they or their family members fell ill with a variety of common ailments, such as colds, the flu, or gastrointestinal problems. While the vast majority used medicinal plants to heal, 41 percent said they also went to the clinic. However, most of this cohort qualified their answers. Some said they only went because they were obliged to take their sick to the clinic, and many complained the clinic was too expensive or that going to the clinic and taking pills did not actually heal their illness.

**Poverty Reduction**

One of the CLAS' original and most important goals was to provide PHC for the poor. In its Letter of Intent to secure loans from the International Monetary Fund (IMF), the Government of Perú (2000) presented the success of the CLAS

---

    4    This is the program under which the CLAS systems falls.
    5    These included doctors, a dentist, medical technicians, and an elected community member.

initiative as proof of its prioritization of social programs and structures as part of its poverty reduction strategy. However, it is important to examine in what ways the CLAS provides a suitable platform for poverty reduction.

Though the state remains in control of their policies and funds the salaries of the doctors and core of health workers (Ewig 2002; Iwami and Petchey 2002), each CLAS is legally a private, not-for-profit entity (Altobelli 2002; Ewig 2002) that has six unpaid community members serving on its board. Half are elected by the community, while the other half are appointed by the head doctor (*gerente*) (Cortez 2000). The CLAS keeps revenues from user fees and prescriptions to pay overhead costs and to reinvest back into the clinic by hiring specialists or purchasing new technologies, thus generating more income. CLAS workers in my sample emphasized that the community being responsible for its own resources is the strongest advantage of the CLAS system. Each CLAS purchases pharmaceuticals from the MINSA in bulk, generic form to keep costs down. Also, like many neoliberal health services, the CLAS charges user fees varying from two to three PEN[6] per consultation with a system in place to protect the poor from having to pay.

That user fees are necessary for the success of a clinic can be a threat to its survival depending on the economic status of the area. Whereas fees accord the clinics more fiscal responsibility and independence, they are an important impediment to one of the CLAS' most publicized goals of providing services for the poor. The doctors' perceptions of the price of user fees, treatments and prescriptions are different than the *campesinos*'. CLAS workers believed they were very affordable; stating a typical visit cost between two and 20 PEN (including prescriptions and tests), with each pill costing only a few cents. One doctor emphasized their affordability was due to state subsidies. Another health care worker insisted that programs and free services supplied by the government convince *campesinos* that the clinic and medicines are affordable. Doctors generally expressed an attitude of benevolence towards the community because of the low cost services they were able to provide.

User fees, however, typically represent approximately 10 to 15 percent or more of a *campesino* family's weekly earnings, if the family has any earnings at all. Furthermore, there was a discrepancy between what doctors claimed a clinic visit cost, and what villagers claimed. *Campesinos* told me it cost between 20 or hundreds of PEN to visit the clinic and do everything the doctors asks. They were aware consultation fees were only two or three PEN, but complained that doctors frequently demand expensive tests, and that pharmaceuticals often must be purchased at a pharmacy because the clinic rarely has them in stock. Doctors did not inform me of these extra costs during the interviews. One local man told me he rarely goes to the clinic because it 'takes money away from the family.' The villagers' capital is often wrapped up in livestock, which they sell when they must access social services or make necessary purchases such as pharmaceuticals.

---

6    Peruvian Nuevo Soles. 1 PEN = 0.34 CAD.

Such a financial coping strategy can be very precarious as some families may be forced to sell livestock in the case of illness, which can send them into a spiral of impoverishment (Janes 1995). For example, one family in Añawi had been economically devastated by the purchase of expensive pharmaceuticals to treat a severe fever that affected many of its members.

As an attempt to counteract high costs of treatment, the CLAS administration argues that profits from user fees gives CLAS the ability to offer free consultation for the neediest (Cortez 2000; Iwami and Petchey 2002). However, as the clinic assumes certain financial obligations, it is not in its interest to provide this service liberally. Furthermore, *campesinos* did not appear to be aware of this option. Only one person from my sample admitted to having her fees exempted. A lively woman in her fifties, she had been gored in the chest by her cow and not having money to pay the doctor, tried to pack the wound with medicinal plants. When her pain was not relieved, she made the trek to the clinic. After being refused treatment because she had no money, she threatened the doctor she would die right there in his office if he did not treat her. The doctor acquiesced. Thus, while in theory the poor can be exempted from paying clinic fees, it was not widely publicized in the clinics I researched seeing as they must be competitive and run a profit in order to survive.

There is a government-subsidized program available for the extreme poor: *Segura Integral de Salud* – Integral Health Insurance (SIS). It is a one time cost of one PEN and provides basic health services such as emergency attention, hospital stays, medicines, and laboratory analysis. I discovered this insurance plan accidentally. A doctor in Clinic A informed me of SIS in passing, making it appear to be an arrangement his particular clinic had. He did not allot it much importance, I did not enquire further, and no other CLAS worker mentioned it. Only once I looked through my photographs of posters that lined the walls of Clinic B did I notice a poster for the SIS, shown in Figure 6.3.[7] Although I made it clear to doctors and health care workers I was interested in investigating specifically how poor *campesinos* negotiated health care in their clinics, no one drew my attention to the SIS. Furthermore, none of the Añawayans I interviewed mentioned this insurance, and their frequent complaints about the cost of laboratory analysis and medicines imply they were not aware of it. Rural villages beset by poverty surround the town

---

7   This poster differed from many of the other cartoon like posters advertising vaccination campaigns, breastfeeding, etc. such as the poster in Figure 6.4. Printed on the poster is a great deal of Spanish text that may be difficult for Quechua speakers to follow. On the bottom of the poster, there is a note that the SIS assures free health care for those who are in extreme poverty. In addition, one must call a toll free number in Cusco rather than purchase it from the clinic. In Añawi, there was one privately owned phone for the village limiting its access, and *campesinos* were not accustomed to making calls. That the telephone number was the only source of information for the insurance suggests that planners of this program did not consider the possible trepidation *campesinos* may feel calling an office in the city, with little or no experience with telephones.

where the clinic is located. The provision of SIS could potentially jeopardize the financial well being of the clinic, which could explain the poor publicity of the program.

**Figure 6.3  Poster for SIS insurance**
*Source*: Author's photo.

**Figure 6.4  Poster for pneumonia prevention**
*Source*: Author's photo.

Indeed, due to the inability of many rural poor to pay user fees, many CLAS clinics in marginal areas are not faring well financially and receive no emergency financial backing from MINSA (Iwami and Petchey 2002). To illustrate, Altobelli

(2002) offers the example of a CLAS clinic in a poor area in the outskirts of Lima that attempted to solve its financial problems by increasing consultation fees and reducing non-income generating programs by 63 percent. Programs may focus on alcoholism in rural communities, stress management, or spousal abuse and were identified by the Alma-Ata declaration as being important to the overall health of populations. Such programs are typical of comprehensive PHC, focusing on issues affecting quality of life (Altobelli 2002). While increasing user fees and cutting programs may be necessary to keep clinics in business, doing so undermines the original mandate of the CLAS of providing low cost health care tailored to local needs. The reliance on revenues necessitates a population capable of assuming those costs and generating income. It follows that this situation is exacerbated in rural areas, where poverty is just as, or more, rampant as in suburban areas (Altobelli 2002). Interestingly, 69 percent of all CLAS clinics are located in rural areas (Cortez 2000).

Health workers maintained that CLAS clinics were sustainable in neither rural nor poor neighbourhoods. One doctor claimed:

> In small villages, I don't think it [CLAS] would work. Because no matter what, the CLAS will always have its price, and every service offered must be paid for. Why? Because with our profits we purchase equipment and logistical materials for the maintenance of the establishment, and pay the personnel.[8]

Another health worker informed me that the first clinic in the Cusco region to become a CLAS was in a relatively wealthy, urban neighbourhood. Profits gained from this economically diverse cohort allowed for the hiring of specialists, purchase of newer technology and coverage of user fees for those who could not afford treatment. Over and over the CLAS workers I interviewed impressed upon me the infeasibility of a CLAS clinic sustaining itself financially in low income areas. In 2000, 22 percent of the CLAS were located in zones of the lowest income quintile and 40 percent in the second lowest quintile (Cortez 2000), suggesting that many of them could face financial difficulties, and calling into question the long term sustainability of the program.

In my study, both clinics expressed lacking funds for necessary expenses. Clinic B had vehicles that facilitated access to rural villages, and served as ambulances when needed. However, they could often not afford fuel, and had to set up a credit system with the local fuel station. Clinic A is in a town that is the final stop before a major tourist destination. The clinic is expected to serve the hundreds of tourists that pass by every day, as well as 17 surrounding *campesino* villages but is too small for this number of clients. In the cramped conditions of Clinic A, health workers could offer little privacy. Low partitions without curtains provided makeshift cubicles that opened towards the centre of the clinic, and while

---

8 The doctor is referring to personnel other than doctors, nurses and technicians who are paid by MINSA.

I was there, several *campesinos* were asked to expose their buttocks for injections in these cubicles. As one CLAS member put it, *campesinos* want their privacy just as much as any other person. It is perhaps for this reason this clinic had difficulty attracting patients, resulting in less profit with which to build a larger clinic. While the committee was able to purchase land to build a new clinic with its profits, it could not afford to construct the building and lacked governmental support to do so, even though the government maintains in its *Decreto Supremo* (MINSA 1994) that they provide adequate infrastructure for CLAS clinics. The clinic had attempted to secure government grants; however, Perú is highly bureaucratised and the difficulty of navigating demands for papers and forms is well known. Once the funding application was submitted, they were continually told to complete forms from other government bureaus, causing them to narrowly miss application deadlines.

### Community Participation

In light of the gross overspending of government during the 1970s and 1980s, it was thought that surveillance by communities would increase fiscal responsibility and control of resources (Altobelli and Pancorvo 2000; Cortez 2000; Noel 2005). The CLAS system achieves this through community participation, a common form of political decentralization that is thought to empower local peoples (Altobelli and Pancorvo 2000; Arredondo 2003). In fact, decentralizing health services is a popular approach to health care reform (Arredondo and Parada 2001). However, the concept of decentralization is very complex and approaches are understood differently by different people (Brohman 1996). There appear to be two main perspectives: administrative and political decentralization (Samoff 1990). Political decentralization is characterized by a focus on the empowerment of populations who are marginalized to make their own decisions (Samoff 1990). Administrative decentralization is characterized by institutional restructuring, a focus on efficiency of services, and often takes responsibility away from the government in order to pay external debts.

By creating a health care structure that includes community participation, the CLAS is potentially able to break from centralized control of the management of health services (Cortez 2000) and move towards a more locally appropriate approach. Accordingly, the CLAS system is reported to be built on a tradition of grassroots organizations that have a history in Perú and is often promoted as a *ground up* response to the demands of locals (Iwami and Petchey 2002).[9] Indeed, elected community participants of the CLAS are entitled to, and have the responsibility to

---

9    It is important to note, however, that this contrasts sharply with Cortez's (2000) account that CLAS is a government initiative.

make decisions not only about functional duties, but also about health programs and promotions.[10] Such characteristics make it appear politically decentralized.

Additionally, there is tremendous opportunity for elected members to bridge differing ideologies of biomedicine and indigenous medicine through community participation. An elected community participant at Clinic A was a *mestiza* who was also governor of the town.[11] She had a great deal of respect for local beliefs, and though she claimed to be largely ignorant of specific local knowledge, her comments revealed she was well informed of local people and their ways. She appeared to be intimately involved with the clinic, and I saw her on several occasions chatting with health care workers about various issues. In one case there was a woman in a remote village that was in danger of having complications during childbirth, but did not want to come to the clinic. Patiently she explained to a CLAS worker why the woman may be frightened to visit the clinic and could also not leave her other children at home alone. Together they discussed a solution to this problem, based on the elected community member's cultural understanding of the woman in question. This incident effectively demonstrates the potential for elected community members to bridge differing ideologies and enhance cultural sensitivity.

However, the model of community participation adopted by the CLAS system is viewed as a shortcoming (Altobelli 2002). There are various ways to obtain community participation (Wayland 2002). In the 1960s when community participation was introduced in the health sector, one of the main reasons it failed to make any lasting changes is because it was often implemented through top-down initiatives legislated by governments (Altobelli 2002). In fact, before the establishment of the CLAS in Perú, community participation referred to vaccination campaigns, which were a decidedly *vertical* (state created) type of community participation (Bowyer 2004). Also, the World Bank has promoted the implementation of user fees (also used in the CLAS system) as a form of community participation (Altobelli 2002), when in fact it is a reform that serves to render health care inaccessible to many.

Before the CLAS, the Peruvian government did not see a need to include community participation in government-established health systems because communities continued to rely on local medical systems including traditional healers, local knowledge, traditional birth attendants and bonesetters (Bowyer 2004). Consequently, when PHC was introduced, vertically appointed committees replaced many traditional social networks that were already in place (Altobelli 2002). Traditional values were considered an impediment to improving the health of the poor, who were perceived to be unable to organize themselves (Ugalde 1985). Additionally, cultural context was not considered in

---

10   By functional duties, I refer to decisions on how to spend profits, who to hire as specialists, and how to allocate funds for various overhead items.

11   *Mestizo* (mixed) usually refers to a person of indigenous ancestry on either side of their family that speaks Spanish and participates in the market economy (Price 2003).

health education, which ultimately limited the exchange of ideas between locals and the state (Altobelli 2002).

While the CLAS has a policy of local empowerment through community participation, it is described as *administratively* decentralized. Thus, the community is expected to participate by overseeing administrative activities such as the allocation of resources, bookkeeping and accounting, purchase of materials and supplies, and communication (Cortez 2000), and if financial circumstances permit they also have input regarding the hiring of specialists. Often, community members lack the education needed to conduct administrative duties (Iwami and Petchey 2002), a significant barrier to their participation in this manner. This is further compounded by the lack of logistical support and guidance from MINSA regarding administrative duties (Cortez 2000). Considering that the rate and quality of community participation depends on each individual community and its capacities (Altobelli 2002), it follows that participation in many of the communities is severely compromised. Indeed, according to all CLAS members interviewed, the most significant barrier in the CLAS system is the low level of community participation.

CLAS clinics are required to act within guidelines decided on by the state and are held to this obligation by virtue of the fact that each clinic depends on the state for three-year contracts (Iwami and Petchey 2002). However, while the state can cancel the contract with the community at any time they esteem that obligations are not being carried out (MINSA 1994) there is no possibility for the community to cancel the contract with the state. Also, important decisions such as the nature of health promotion programs and organization of the CLAS system are made centrally either in Lima or regional health offices, while the community and clinic merely control fiscal and administrational duties. Administrative decentralization is often implemented in order to increase efficiency of services at the expense of empowerment (Samoff 1990). In fact, decentralization of this nature is better understood as a 'façade for maintaining or reinforcing central authority', thereby 'reinforcing power relations' (Samoff 1990, 517). Although community participation is presented as a high priority of CLAS, it is actually subordinate to (fiscal) efficiency (Cortez 2000; Iwami and Petchey 2002) suggesting that the government, not the community, retains control over their clinics.

Moreover, sacrificing meaningful community participation for the benefit of fiscal efficiency comes at an expense. Consequently, disenchantment abounds both with community and health workers. The CLAS system has recently increased its demands on doctors. While they are obligated to work a minimum number of hours, there is no maximum number of hours they can work. My observations suggest that the government profits from doctors' commitment and sense of responsibility and health workers were encouraged to work as many hours as it took to make their clinic (financially) successful. CLAS positions are competitive due to limited job opportunities for trained health professionals, and health workers undergo evaluation by the community every three months. Thus, doctors do everything in their power to keep their appointments, even if it means lack of sleep. The *gerentes*

I interviewed demonstrated strong commitment to improving their clinics. For example, the *gerente* of Clinic B works approximately 700 km (or an 18 hour bus ride through the Andes) away from his hometown where his wife and children reside. He visits them once a month, and after working a ten hour shift, spends most nights sleeping in the hospital in case of an emergency. Thus, a commonly touted benefit of a clinic being administered by CLAS was that there was someone on staff 24 hours a day. However, in both clinics, the *gerentes* claimed to be overworked and visibly fatigued. In such a state it is debatable whether doctors can perform at their best.

The *gerente* of Clinic B regretted there was no one at his administrative and medical level to verify his actions and help him avoid mistakes due to fatigue. In fact, both *gerentes* informed me they often assumed administrative duties that community members were unable to complete. Consequently, extra time spent to ensure the absence of errors exacerbated their exhaustion. In urban CLAS clinics, this is less pronounced as the elected community members are often more educated than their rural counterparts, and are better prepared to assist in administrative duties. The resulting lack of community participation in rural areas causes the frustration of locals, health workers and the state. The image fostered through the CLAS initiative, that the state cares about the health of its citizens, is being undermined due to the narrow scope of community participation (Bowyer 2004). This is exacerbated by past breaches of trust that are still alive in people's memories, which include threats from doctors that legal problems would result if locals refrained from bringing their sick to the clinics. As is characteristic of administrative decentralization, power relations from the past are reinforced and continue to colour interactions between the clinic and locals, revealing that local empowerment is not being achieved

Furthermore, it has been reported that only approximately 20 percent of the population is aware that community members are elected (Iwami and Petchey 2002) and that CLAS is jointly administered by the state and community. Bowyer (2004) reported that in regions CLAS clinics cover, 95 percent of the locals did not participate in electing community members and 91 percent did not know who elected members were. In fact, it is not clear how the democratic election of community members is undertaken. Additionally, many of the elected CLAS members are not fully knowledgeable of the institutional structure and processes of the CLAS and thus have very limited social control and authority over policies and programs (Altobelli 2002). My research corroborates these findings. *Campesinos* I interviewed did not recognize the difference between state-sponsored clinics and CLAS clinics, and did not know how or when to vote for community participants. The doctors I interviewed agreed that most people in the community were unaware that community participants were elected.

It has also been reported that in some communities, CLAS participants are appointed rather than elected (Altobelli 2002). CLAS workers I interviewed indicated this was the case, or that they were local elites, reifying power relations. Some community participants actually lived outside the area, and often did not attend obligatory monthly meetings, suggesting they were disconnected from locals. One doctor suggested that many elected members enjoyed the prestige that comes with this position, but were not interested in the clinic. In Clinic B, elected members were never present, and doctors were unwilling to put me in contact with them, claiming their only participation was during monthly meetings where health workers report to community members. I suspect they did not live in the village. The lack of meaningful participation led one doctor in the clinic to refer to community participants as playing an honorary role. This puts into question CLAS claims of grassroots organizing and empowering local community members.

While Cortez (2000) claims that community participation in the CLAS can at best be characterized as 'active but dependent', my research suggests it is debatable how active the community was at all. Rather than being part of the process of improving the health sector, the type of community participation embodied in the CLAS acts as a means to an end by offering the government a way to fulfill one of the principles of comprehensive PHC (Altobelli 2002). The suggestion that the CLAS system often pays lip service to community participation (Iwami and Petchey 2002) is corroborated in this research project, with the important exception of one elected member. While this participant was extraordinary in her commitment to the clinic, for my sample she was the exception rather than the norm. Furthermore, the question remains as to whether her participation can by categorized as *grassroots* and *local* given that she is the governing official of the town. It is no question that she was an asset to the clinic; however, given her position of power over the local indigenous people it is unknown what her relationship with them was. Other elected members were referred to as occupying no more than an honorary role, as their activities consisted of nothing more than attending a two-hour monthly meeting.

## Conclusion

In this chapter, I have discussed the CLAS system from the perspective of both health care workers and *campesinos* in a rural community in order to examine two goals of this PHC system in Perú: poverty reduction and community participation. Despite having comprehensive goals, the CLAS is more characteristic of the selective approach to PHC, and appears to prioritize (fiscal) efficiency over equity. The CLAS is organized as a private health system and relies on user fees from patients to finance all costs of the clinic except for the salaries of essential health care workers.

The implementation of user fees has marginalized low-income sectors of Peruvian society. Regardless of how inexpensive the user fees appear to be to CLAS workers, they represent a significant percentage of the insignificant (and sometimes non-existent) earnings of poor residents. CLAS cannot claim to be part of a poverty reduction strategy when poor *campesinos* must sell their animals or go into debt in order to access services. Also, the CLAS has not met its goal of increasing coverage for low-income rural sectors. While CLAS clinics function in areas of either high or mixed income, in rural areas, where most residents are poor, the CLAS is not financially sustainable. It follows that in such areas the overt promotion of the SIS insurance and the exoneration of fees are not in the clinic's interest.

While CLAS promotes decentralization as an empowerment tool for local communities, decisions concerning health programs and promotions continue to be conducted centrally. Additionally, community participation is limited to administrative duties. While this can function in urban settings where the population is more educated, this is rarely the case for rural *campesinos*. Thus, community elites often occupy this role, thereby reinforcing power relations within the community, or occupying only an honorary role. This is neither beneficial to the community nor the clinics, since the managing doctor must fulfil this role in addition to his or her regular duties. However, there are a number of other duties that locals could engage in: translating, welcoming and admitting patients, and providing indigenous health care or cultural knowledge. The position of 'elected community participant' would be more meaningful if the position were remunerated at a level that reflects local community members' incomes. In Añawi, I was told that the local wage for a day's work outside the village was approximately seven to ten PEN (CAN \$2.40–3.40). A comparable wage may not attract local elites, but may attract local *campesinos*. Remuneration may also lead to more meaningful community participation in which community members are actively and practically involved in providing health services. Future research should explore the feasibility of such endeavours.

It is likely that the CLAS and other expressions of health sector reform will remain the primary source of PHC in the near future (Hall and Taylor 2003). Therefore, the CLAS should be seen for what it is, a medical enterprise driven by financial and commercial imperatives that attempts to maintain its authority through control and power relations (Miles 2003; Wayland 2003). Therefore, it should not claim to constitute a comprehensive approach to PHC. In fact, in rural areas, the CLAS system operates more like a selective health system for the rural poor who are more likely to look after PHC in their own homes and communities.

**References**

Altobelli, L.C. (2002), 'Participación comunitaria en la salud: La experiencia peruana en los CLAS', in J. Arroyo (ed.), *La salud peruana en el siglo XXI: Retos y propuestas de política* (Lima, Perú: Consorcio de Investigación Económica y Social).

Altobelli, L.C. and Pancorvo, J. (May 24–26, 2000), *Shared Administration Program and Local Health Administration Associations (CLAS) in Peru.* Paper presented at the Challenge of Health Reform: Reaching the Poor. Europe and the Americas Forum on Health Sector Reform, San Jose, Costa Rica.

Arredondo, A. and Parada, I. (2001), 'Financing Indicators of Health Care Decentralization in Latin America: Information and Suggestions for Health Planning', *International Journal of Health Planning and Management* 16, 259–76.

Arredondo, H.C. (2003), 'Los Comités Locales de Administración en Salud (CLAS) ¿Buenos o malos para la salud?' *Situa – Revista Semestral de la Facultad de Medicina Humana – UNSAAC* 12:25, 54.

Asthana, S. (1994), 'Economic Crisis, Adjustment and the Impact on Health', in Phillips, D.R. and Verhasselt, Y. (eds), *Health and Development* (London and New York: Routledge).

Bowyer, T. (2004), 'Popular Participation and the State: Democratizing the Health Sector in Rural Perú', *International Journal of Health Planning and Management* 19, 131–61.

Brohman, J. (1996), *Popular Development: Rethinking the Theory and Practice of Development* (Massachusetts: Blackwell).

Cortez, R. (2000), *La entrega de servicios de salud en los centros de administración compartida (CLAS): El caso del Perú* (Lima: Centro de Investigaciones de la Universidad del Pacífico).

Cueto, M. (2004), 'The Origins of Primary Health Care and Selective Primary Health Care', *American Journal of Public Health* 94:11, 1864–74.

Ewig, C. (2002), 'The Politics of Health Sector Reform in Perú', *Woodrow Wilson Center Workshops on the Politics of Education and Health Reforms*. Washington DC, April 18–19, 2002.

Hall, J.J. and Taylor, R. (2003), 'Health for All Beyond 2000: The Demise of the Alma-Ata Declaration and Primary Health Care in Developing Countries', *Medical Journal of Australia* 178:1, 17–20.

Herbert, S. (2000), 'For Ethnography', *Progress in Human Geography* 24:4, 550–68.

Homedes, N. and Ugalde, A. (2005), 'Why Neoliberal Health Reforms have Failed in Latin America', *Health Policy* 71, 83–96.

Iwami, M. and Petchy, R. (2002), 'A CLAS Act? Community-based Organizations, Health Service Decentralization and Primary Care Development in Peru', *Journal of Public Health Medicine* 24:4, 246–51.

Janes, C.R. (1995), 'The Transformation of Tibetan Medicine', *Medical Anthropology Quarterly* 9, 6–39.

Kearns, R. (2000), 'Being There: Research Through Observing and Participating', in I. Hay (ed.), *Qualitative Research Methods in Human Geography* (Victoria, Australia: Oxford University Press), 103–121.

Kearns, R. and Moon, G. (2002), 'From Medical to Health Geography: Novelty, Place and Theory After a Decade of Change', *Progress in Human Geography* 26:5, 605–25.

Londoño, J.L. and Frenk, J. (1997), 'Structured Pluralism: Towards an Innovative Model for Health System Reform in Latin America', *Health Policy* 41, 1–36.

Madge, C. (1998), 'Therapeutic Landscapes of the Jola, The Gambia, West Africa', *Health & Place* 4:4, 293–311.

Miles, A. (2003), 'Healers as Entrepreneurs: Constructing an Image of Legitimized Potency in Urban Ecuador', in J. Koss-Chioino, T. Leatherman and C. Greenway (eds), *Medical Pluralism in the Andes* (London and New York: Routledge), 107–28.

Noel, R.S. (2005), 'Compartiendo las responsibilidades', *IDBAmérica, Magazine of the Inter-American Development Bank*, April 13, 2005.

Price, L.J. (2003), 'Illness Management, Social Alliance, and Cultural Identity in Quito, Ecuador', in J. Koss-Chioino, T. Leatherman and C. Greenway (eds), *Medical Pluralism in the Andes* (London and New York: Routledge), 209–33.

Samoff, J. (1990), 'Decentralization: The Politics of Interventionism', *Development and Change* 21, 513–30.

Ugalde, A. (1985), 'Ideological Dimensions of Community Participation in Latin American Health Programs', *Social Science & Medicine* 21:1, 41–53.

Wayland, C. (2002), 'Acceptable and Appropriate: Program Priorities vs. Felt Needs in a CHW Program', *Critical Public Health* 12:4, 335–50.

Wayland, C. (2003), 'Contextualizing the Politics of Knowledge: Physicians' Attitudes Toward Medicinal Plants', *Medical Anthropology Quarterly* 17:4, 483–500.


**Internet-based References**


Future Generations (2004), Perú: Country Program Overview. *Future Generations.* Available at <http://www.future.org/pages/03_country_programs/3_peru/00_peru_intro.html>, accessed May 19, 2005.

Government of Perú (2000), Letter of Intent. *Written to the IMF.* Available at <http://www.imf.org/external/np/loi/2000/per/01/index.htm>, accessed May 19, 2005.

Ministerio de Salud (MINSA) (1994), *Decreto Supremo N 01–94–SA.* Available at <http://www.paag.minsa.gob.pe/internet/QueHacemos/clas/documentos/Dec retoSupremo0194SA.pdf>, accessed September 25, 2006.

Ministerio de Salud (MINSA) (2004), *Mortalidad Materna.* Available at <http://www.minsa.gob.pe/portal/03Estrategias-Nacionales/05ESN-SaludSexual/ss-sr.asp>, accessed September 25, 2006.

# PART 2
## People

Chapter 7

# Geographies of Family Medicine: Describing the Family Doctor's Practice-Based Landscape of Care

Gina Agarwal

The purpose of this chapter is to explicate the implicitly and explicitly geographic elements of the family doctor's practice. These elements contribute to a larger landscape of care operating within the discipline of family medicine. Such explication is necessary because there is scant literature examining the landscape of care within which family doctors operate, despite the fact that there are many spatial elements inherent to the practice of family medicine. An overall objective is to illuminate the complex way in which the structure and nature of multiple spaces – the community, clinic, and consultation room, among others – have implications for how family doctors undertake their clinical work and develop relationships with patients in their roles as core providers of frontline care guided by the primary health care (PHC) mandate. Examples of these spatial implications are drawn out, based primarily on family medicine practice in Canada and the United Kingdom (UK).

In order to explore the geographies of family medicine, it is important to first understand what PHC is and the roles of the general practitioner (GP) within it. PHC is the main mechanism for health care delivery at the most local level (frontline) of a country's health system. It consists of basic and essential health care that the immediate community has access to. It encompasses treatment of common diseases or injuries, provision of essential drugs, maternal and child health, prevention and control of locally endemic diseases and immunization, community health education about common health problems and how to prevent them, and the promotion of an adequate diet and lifestyle. Primary care is the first tier in most health care systems and is typically the first point of contact patients have with formal care; it is the tier most guided by the PHC mandate. It represents approximately 56 percent of all health care services accessed daily in the province of Ontario, Canada (Manuel et al. 2006, 5). Family practitioners, family physicians or GPs are doctors that work within the primary care system, and under the PHC mandate, as 'frontline' practitioners. They are thus central to a PHC framework as their practice includes many frontline care activities. They also often work in collaboration with other primary care providers such as nurses, social workers, pharmacists, dietitians, physiotherapists and occupational therapists.

Unlike specialists, GPs typically do not limit themselves to caring for a specific illness or body system, but are available for any health problem. They have diverse and diffuse roles, which are often complex and draw upon multiple skills, often simultaneously. The European Academy of Teachers in General Practice (Euract 2005, 6) considers the responsibilities of GPs to be: '... the provision of comprehensive and continuing care to every individual seeking medical care irrespective of age, sex and illness. They care for individuals in the context of their family, their community, and their culture ... They recognize they will also have a professional responsibility to their community ...' In short, GPs can be seen as frontline practitioners who are intimately connected to the environment and community in which they practice (Manuel et al. 2006). Here environment constitutes the proximal elements, factors and conditions which may have an impact on the development, actions or lifestyle of a person. Community can be defined as 'a group of people with diverse characteristics who are linked by social ties, share common perspectives, and engage in joint action in geographical locations or settings' (MacQueen et al. 2001).

That GPs are typically the first place or point of contact for patients entering the first tier of health care implies that they are not only physically co-present with patients, by the fact that they have offices which are local, but also that they practice in the same community and environment in which their patients live and work. Further, their offices are often a visible part of the community. Throughout the world GPs can be accessed in different places (e.g., clinic, hospital, remotely via telemedicine or Internet technologies, travelling clinics, at work, in a care recipient's home), but their role as a primary care practitioner stays relatively constant.

In Canada, the *Four Principles of Family Medicine* (College of Family Physicians of Canada 2003) succinctly summarize the different aspects of family medicine. These principles are applicable across all types of family practice (i.e., across care settings and patient groups). In this chapter I use these principles as a structure upon which to base a further discussion of what I refer to as the geographies of family medicine, ultimately working toward addressing the purpose set out above. I specifically illustrate the nature of modern family medicine with respect specifically to notions of place. In each subsection I first introduce a specific principle and then follow it with further discussion and explanation, with relevant examples. Two of the four principles most explicitly involve the concepts of community and environment; these are discussed first. The chapter is concluded by reflecting on the spatiality of my own practice as a GP in the city of Hamilton, Canada.

## Family Medicine is a Community-based Discipline

The family physician may care for patients in the office, the hospital (including the emergency department), other health care facilities, or the home ... [they] see themselves as part of a community network of health care providers and are skilled at collaborating as team members or team leaders (College of Family Physicians of Canada 2003).

The principle described above captures the notion that the work site of the GP extends beyond the boundaries of her or his physical clinic space. Family practice is based in the community and is significantly influenced by local factors. In this way, many GPs live and work locally and share these community-based ties with other members who, in turn, become their patients. They use the same shops and services as patients and are also exposed to the same social, political, cultural and linguistic environments. Furthermore, as part of their practice, GPs become knowledgeable about the facilities and resources available in the community and to different community groups. GPs, thus, must be able to absorb a repertoire of essential information regarding the community in which they practice and construct 'mental maps' about the locality (Holden 2004). For example, they should come to know locations of care facilities, amenities and support groups, public transportation availability, areas of socio-economic deprivation, and cultural-linguistic groups within the local area and key spaces of residents' lives (e.g., workplaces). In situations where GPs are not resident in the community in which they work, they may become knowledgeable about the local environment over time and/or with the assistance of clinic staff.

It is understood that GPs need to be on-site in a community for at least six months before they have a mental map of local health and social services useful enough to be drawn upon in consultations (Holden 2004). After completion of training, GPs typically access their patients based specifically on where they locate their practice and the catchments around the clinic site. Location can affect the availability of potential clinic sites and also shape the environment with respect to demand for services (i.e., whether there is a need for a GP in the community or not), local and regional determinants of health (e.g., Evans and Stoddart 1990), and even quality of work life for the practitioner. Such nuances must be learned through immersion in the local environment for long enough to be able to both create and use the mental map.

GPs act as co-coordinators of patients' care and typically use their knowledge of the community in doing so. They also serve as 'gatekeepers' to secondary and tertiary care. GPs, by virtue of the fact that they are the first point of contact that patients have with the health care system, will largely determine which referrals are made into the second and third levels of care (McWhinney 1997). An outcome if this is that they play some role in determining which local *and* non-local health services patients in need of care will ultimately have access to or utilize. Because of this they must also be aware of health service providers, particularly specialists, who are available to members of the local community but do not practice within it.

Because the GP provides services for the community and also is a community member, it is important that s/he practices with the best interests of the community in mind. In some instances, however, there can be ethical dilemmas for the GP in that the best interests of the patient may not be in keeping with the best interests of the community. A common example of such a dilemma comes from recognizing that having patients use particular resources or services can both help to improve the health of the individual patient while also lessening access to such care for others in the community due to limited availability. Policy changes have sought to make GPs more accountable for how they recommend services and allocate resources, with the goal of distributing resources more equitably among community groups and individuals in need. Increasingly, GPs accept responsibility for the stewardship of scarce resources while simultaneously considering the needs of patients and the community (Rosser 2006). In this way they are given a role in maintaining a care infrastructure for their patients.

Patients can be seen in their homes, hospitals or other places (McWhinney 1997) and family physicians must adapt their skills in order to be able to offer care in different settings. They must understand and negotiate each new care environment that is presented to them. For example, both Parr (2000) and Philo (1997) contend that the stigma associated with mental illness in the Scottish Highlands results in the 'giving and receiving [of] care' being a contested practice. In such an instance, this stigma serves as a powerful element of the social environment of care within the community and can ultimately impact upon a GP's ability to practice. Understanding the nuances of particular care environments (whether in the community, institutional or private settings) and also local attitudes towards care are thus central to a GP's practice.

Medical care offered by the GP in a patient's home requires overcoming established boundaries of privacy (Conradson 2003). However, the contextual information that a GP can gather from visiting a patient's home is very rich and thus it might be useful to overcome such boundaries and in doing so risk medicalizing the home. Clues about the patient's lifestyle, family and social context are gained which can help to identify a diagnosis, confirm a risky behaviour, or start an appropriate management program. In such instances, the clinician's understanding of a patient's place of habitation is very important. The GP will have increased understanding of the patient's social and environmental situation, potentially leading to a more holistic approach to medical care being employed (Holden 2004). A holistic approach is one in which the GP considers the whole patient in his/her consultation. The term 'whole patient' implies simultaneously caring for the patient's physical, psychological, social and emotional well being. By knowing the social context from having conducted a home visit, a patient's headache, for example, can assume a totally different diagnosis (e.g., serious and life threatening cerebral hemorrhage versus common and treatable tension headache). Increasingly GPs are challenged to find the time to undertake home visits due to increasing responsibilities and job expectations, an outcome of which is a lessening of their abilities to understand the dynamics of patients' homelife.

Though community and face-to-face medicine have their place, nowadays patients are consulted over the telephone more and more (Manuel et al. 2006) and in some cases e-mail communication is gaining in popularity. In other instances, GPs may be flown in to hold special, temporally-limited clinics for residents of rural, remote, or generally underserviced areas, or may be involved in mobile clinics that serve large geographic areas. In general, remote consultations in the form of internet-based communication and telemedicine, mobile clinics, and temporally-limited (non-local) practice all challenge the notion that the GP is a member of the community. Because of the importance of understanding the local community, GPs practicing under such circumstances must work to overcome their lack of (lived and accumulated) knowledge about the local environment. In the case of GPs involved in traveling clinics or temporally-limited non-local practice, they may address this challenge through utilizing the knowledge of local people, which could involve asking people explicitly about the community in which they live, and through obtaining information from more experienced medical staff on-site. The continuous nature of patients' records may also serve to assist these GPs in locating community resources.

In countries with large rural populations such as Canada, communities and the nature of the local environment can significantly differ across the urban-rural-remote continuum. Rural family practice requires very different skill sets from urban family practice. For example, GPs in rural practice are more likely to be involved in providing palliative care than those practicing in urban centres (CIHI 2005). Rural GPs need to be more versed in acute and emergency medicine, utilizing specialists only when they have to due to particularly significant geographic barriers to access, and in working as a team with fewer practitioners. Urban GPs will refer patients more freely, and can take advantage of the greater array of resources available in metropolitan centres. Recruitment and retention of rural family physicians is also a significant problem (Anderson et al. 1990; Coyte et al. 1997). This has resulted in Canada's College of Family Practice making an active effort to attract more medical students to this style of practice (Krupa and Chan 2005). The consideration of differences in a GP's practice between rural and urban places assists in making Kearns and Moon's (2002, 610) point that ' … places matter with regard to health, disease and health care.'

### The Family Physician is a Resource to a Defined Practice Population

> The family physician views his or her practice as a 'population at risk', and organizes the practice to ensure that patients' health is maintained whether or not they are visiting the office … [and must] advocate for public policy that promotes their patients' health (College of Family Physicians of Canada 2003).

In the context of family medicine, 'population' refers simultaneously to multiple groups. This term references not just the patient population of the practice, but also the members of the community at large, as well as the various social groups (e.g., ethnicity, gender, age, sex, occupation) to which the patients affiliated with the practice belong (Parchman and Burge 2002). The notion of the 'population at risk,' stated in the above description, refers to the fact that the GP must constantly recognize that the entire practice population – if not the population at large – is at risk of becoming ill and that reducing risk is as central to his/her practice (e.g., through delivering immunizations) as treating illness.

To meet the needs of the practice population, GPs have an ability to alter the types and modes of practice they employ. In areas of social deprivation, for example, practitioners must be more aware of the economic and social factors that may affect patients' adherence to treatment (e.g., due to the high cost of pharmaceuticals) and/or abilities to keep follow-up appointments (e.g., due to an inability to access affordable transportation). Particular communities will be burdened by illnesses not prevalent in other areas. For example, colposcopy screening – a procedure involving a visual examination of the cervix and surrounding tissue – may be more needed in inner-city areas since the population in such areas is at higher risk of cervical cancer. This is because such areas are typically comprised of younger women of lower socio-economic status and larger populations of sex industry workers, these being groups at risk of developing cervical cancer. However, at risk patients may have little funds to pay for transport and/or find it difficult to take time off work or pay for childcare in order to attend colposcopic appointments in specialists' offices, typically located in hospitals or large medical centres not situated in the inner-city. In one venture, an inner-city community health centre opened its own colposcopy screening unit (Ogilvie et al. 2004). The GP was trained to perform diagnostic colposcopy (but not to provide surgical treatment). The program was successful in that less people defaulted on their appointments in the community health centre clinic compared to the hospital clinic (1.3 percent versus 17.1 percent, p<0.01). This is a successful example of a clinic and its GPs reshaping service delivery in order to meet the specific needs of the practice population.

Some GPs do not exclusively treat rostered patients from a defined practice area. In socio-economically deprived and under-serviced areas there are large numbers of individuals who do not have regular GPs and are at risk of becoming ill and/or not receiving needed care, particularly highly marginalized groups such as those experiencing addiction, homelessness, and mental illness. In fact, such marginalized groups are at greater risk of not being able to obtain necessary PHC (Evans, Barer and Marmot 1992) and are unlikely to seek/use any form of PHC regularly. In response, some GPs work in local shelters or with charitable organizations that provide mobile health clinics and as such do not have a set roster of patients. Others work in mobile health vans. In the city of Hamilton, Ontario such a mobile health van program has been established to provide PHC to different street populations (City of Hamilton 2007). Other GPs work in satellite PHC clinics that serve difficult-to-reach populations but have a main base in larger primary care clinics.

GPs may actively choose to practice in a specific location due to the dynamics of the surrounding population, both at-large and in the clinic. GPs are well aware of the fact that illnesses have different prevalence rates in different locations and that different communities have access to different resources, and as such will choose to work in areas that interest them professionally. For example, GPs who enjoy working with HIV+ patients will chose to work in areas where infection rates are higher and where resources needed by this specific population are provided in the form of HIV specialist clinics and anti-retroviral drug coverage.

Another fundamental part of family medicine practice is the provision of preventive medicine and health promotion. Indeed, as mentioned in the introduction, prevention of disease and promotion of a healthy lifestyle are part of the primary care health care structure and providing such care is also in keeping with the PHC mandate. Health promotion and preventive medicine must be tailored to the practice population through understanding the needs patients may have and the risks they may encounter in their local environments. GPs, by knowing patient records and patients' clinical and personal backgrounds, will seek to offer regular preventive medical services (such as immunizations, cervical screening, mammography screening, cardiovascular health assessment, and diabetes screening) to ensure maintenance of good health. This requires meticulous record keeping and ensuring that certain screening tests are not lost to follow up and also requires that patients receive appropriate checks in a timely fashion (Aaronson et al. 2001).

Further to the above point, GPs offer two different methods of health promotion: (1) *targeted* (to a population of high risk – for example, certain ethnic groups have a high risk of developing diabetes and these individuals should be screened frequently); and (2) *opportunistic* (when a patient attends the clinic for another reason and an opportunity for health promotion arises). An example of a targeted community health promotion intervention for diabetes screening is the Community Health Awareness of Diabetes (CHAD) project (Agarwal and Kaczorowski 2006). In this instance, screening was taken out of the GP's office and placed into local pharmacies in an effort to attract people who may not regularly visit a clinic. This was done to help GPs with their ever-increasing workload and to make sure that screening of individuals at high risk for developing diabetes actually took place. The program assessed large numbers of people for diabetes risk in the community, but sent the results of the risk assessments to their own GPs so that they would formally screen those at high risk for developing diabetes. This is an example of a health promoting initiative designed to take screening out of the GP's office and into the community, not just for case-finding purposes but to increase community awareness of certain diseases within the population at large.

## The Family Physician is a Skilled Clinician

> Family physicians demonstrate competence in the patient-centred clinical method; they integrate a sensitive, skillful, and appropriate search for disease ... [and] have an expert knowledge of the wide range of common problems of patients in the community, and of less common, but life threatening and treatable emergencies in patients in all age groups (College of Family Physicians of Canada 2003).

This first point described above refers to the competence of the family physician with what can in part be described as the 'clinical geography' of consultations. Here, the clinical geography can be defined as the layout and nature of the medical, psychological and social details of the patient and how they unfold themselves within the context of the consultation, as well as the process measures of the consultation itself and at what point they are reached. Process measures within a consultation refer to core aspects that must be acted out, such as a greeting at the beginning, a session listening to the patient's problem, a summary back to the patient of what the problem is, a detailed history taking or information gathering followed by a clinical examination, and then a negotiation of further management and follow up with the patient. Indeed, as a learning exercise for GPs in training, consultation mapping has been encouraged (Arborelius and Bremberg 1992). This is a method of visually laying out the components of the consultation (i.e., the process measures) in relation to time. The visual map – which aesthetically appears as a chart with one axis representing time and the other specific process measures – will serve to classify the consultation type and allow the cartographer, in this instance the GP, to see what aspects may have been missed or left out of the consultation and put them under scrutiny. A skill of GPs, then, is the ability to recognize and map out the clinical geography of each consultation, with each map covering relevant spatio-temporal issues central to the consultation.

The GP typically develops a relationship with the patient before deciding – based on skill, knowledge, evidence, and other important factors – what illness or condition a patient is experiencing, unlike specialists who often must treat patients with whom they have little established relationship. The space in which each particular consultation occurs is of central importance to the relationship-building process. This is because the space of care provision plays a role in shaping the consultation between a GP and his/her patient and its outcomes. Crooks (2006, 53) notes, for example, that 'the space of care provision is the place in which the doctor-patient interaction occurs; it is shaped by larger forces, such as the organization of the health care system and the social construction of lay and professional knowledge, and micro forces, such as one's intersectional performance as a … patient.' From this definition we can understand that part of a GP's skilled practice must be recognizing those factors that ultimately shape whether and how s/he is able to develop a meaningful and even potentially therapeutic relationship with patients.

The temporal and spatial contexts of interactions between GP and patient assist the practitioner in piecing together the diagnostic puzzle and even with coming to a diagnosis (Euract 2005), this being a specific clinical skill of the GP. Recognition of diagnosis as a process that takes place over space and time that is influenced by variables both within and beyond the space of care provision is important in the diagnosis of psychological and mental illnesses as well as chronic diseases in particular. GPs often seek spatio-temporal distance from the patient and his/her symptom(s) in order to observe a multiplicity of factors (e.g., changes in symptoms over time, lifestyle influences) as a way to confirm the diagnosis beyond further doubt (Euract 2005). Specifically, the spatio-temporal distance afforded between consultations in which information is gathered can allow the diagnosis to become clearer and more formulated.

Social geography at a wider scale, beyond that of the clinic setting, is also important to the GP. Here social geography can be described as the setting for the patient's illness or problem within the context of his/her familial, societal, economic and political environment and the potential for these environmental factors to act as constraints or enablers to achieving health. As a skilled clinician, the GP must contextualize the patient and his/her illness behaviour in an attempt to understand the gains and losses to an individual's life-in-place as a result of onset (Illich 1976). For example, there may be financial or social losses to be had due to an inability to work after becoming ill. In such an instance, the GP must apply his/her skills, particularly those related to information-sharing and relationship development, in order to gain an understanding of the environmental factors that not only shape the cause(s) of illness but also in relation to its outcomes (McWhinney 1997).

The practice of medicine has always been intimately connected to the physical environment, as reflected by Hippocrates in his comments (400BC): 'When one comes into a city in which he is a stranger, he ought to consider its situation. ... one ought to consider attentively the waters the inhabitants use ... From these things he must proceed to investigate everything else' (translated by Adams, n.p.). Because of GPs' long-standing connections to the environment and community, they make natural epidemiologists and public health consultants. This can be considered another skill of the GP. John Snow, for example, was a local doctor in the mid 1800s in Britain who is recognized as the first epidemiologist. This is because it was through his knowledge of patients, their families, their behaviours, their living environments, and the community that he was able to locate the source of a cholera outbreak in London to a nearby water pump (Vinten-Johansen et al. 2003). This type of sleuth-like investigating and resulting discovery could only have been achieved through his working knowledge of the physical and social geographies of both risk and disease outbreaks in the community. This is in keeping with Pemberton's (1969) contention that the GP can and should be connected to his/her community and use it as a field laboratory for epidemiology or the study of patterns of disease.

## The Patient-Physician Relationship is Central

The patient-physician relationship has the qualities of a covenant – a promise, by physicians, to be faithful to their commitment to patients' well-being, whether or not patients are able to follow through on their commitments. ... They use repeated contacts with patients to build on the patient-physician relationship and to promote the healing power of interactions (College of Family Physicians of Canada 2003).

The doctor-patient relationship is one which is very unique in family medicine; it involves chronologically mapping out the patient's life and is dependent upon the continuous contact between patient and doctor. A GP may have the opportunity and privilege to care for the patient from birth to death or at least for an extended period of time. Popular writers such as A.J. Cronin (1978) have become successful with the popularization of the image of a friendly GP with whom patients develop long-term relationships. In his many books the main character, Dr. Finlay of Tannochbrae, was a charming solo GP who worked in the remote and sparsely populated Scottish Highlands. He was very popular in the community and knew all his patients socially and medically over the years he practiced. His trademark was that he knew his patients from birth to death, thereby offering them continuity of care. Continuity of care is an important outcome of the development of a long-term relationship between GP and patient and is thus implicitly tied to this particular principle of family medicine.

Experts have described continuity of care as the: relationship between patient and doctor beyond single illness episodes; affiliation between doctors and patients; and sense of loyalty and personal continuity that leads to trust, communication, and responsibility (Freeman and Hjortdahl 1997). There are three dimensions of continuity of care: informational (Curtis and Rogers 1980; Hansen 1975; Wall 1981), or the availability of patient information to health care providers; interpersonal/relational (Curtis and Rogers, 1980; Haggerty et al. 2003; Hansen 1975; Hennen 1975), in which the patient repeatedly consults a particular family doctor who has been acquainted with him/her through health episodes for some time; and longitudinal (Haggerty et al. 2003; Hansen 1975; Hennen 1975; Reid, Haggerty and McKendry 2002; Wall 1981), or a patient's repeated use of a single health care site.

Indeed, continuity of care is important because it improves health outcomes (Saultz and Lochner 2005). It can influence patient satisfaction (Reid, Haggerty and McKendry 2002), doctor satisfaction, and patient outcomes (Mainous et al. 2004). It is also associated with a lowered cost of care, improved delivery of preventive services, and lower rates of hospitalization. Continuity of care can also influence the physician's use of time within the consultation (Mechanic 2001), as a patient's regular doctor may spend less time consulting on certain things than an unknown doctor. Longitudinality is important to long-term relationship building between the patient and the doctor. By virtue of having repeated contact with individual patients over time within the clinical space, GPs are able to gain an understanding of each patient's 'family geography' (Holden 2004). Family geography can be defined as the genetic and socio-cultural elements that constitute a patient's whole family. Specifically, the consultation is used as a vehicle for expanding knowledge of the patient and his/her family members to assist in determining directions for illness management and future care. Repeated chronological contact with the patient in various circumstances and at various points of care cements the relationship and trust between the patient and GP. Importantly, having a well-known and continuously visited doctor can actually facilitate improvement in health status as a result of the development of a therapeutic relationship (Balint 1972).

It is essential that GPs be aware of how their own personality, training, and social geographies (i.e. their positionality) shape their clinical skills and understandings of their patients' lives and experiences of health and illness. Recognition of this, in both training and subsequent clinical practice, allows GPs to overcome potential differences and empathize with their patients. GPs can use different strategies to maintain closeness to and empathy for their patients without necessarily being socially or economically proximal. For example, they may acknowledge major life events in the families of their patients by a making a phone call or mailing a card. They may visit their patients in hospital in order to play a part in their care in other settings (e.g., a long-term care residence). They may attend funerals of their patients when appropriate. They often return phone calls from patients when they are worried or concerned and in need of medical advice or assistance. It is thus important to build trust between GP and patient over time so that at specific times of medical need the doctor can act as a support and resource and engage in more empathetic practice.

## Concluding Discussion: Reflecting on My Own Practice

In concluding this chapter I now offer some personal reflections regarding my experience as a GP in clinical family practice with reference to the issues that have been discussed at length above. In my practice over the years, space and location have played key roles. They have affected how I interact with my patients and how my patients interact with me. When previously practicing in urban areas in the UK, undertaking home visits was more prevalent than in my current practice in urban Canada. The home visits I made connected me with my patients and gave me an intimate knowledge of what made certain patients live and become ill in the ways they did. As time progresses, my use of the internet and phone as a method of consulting my patients is becoming more common and useful. In Canada, I can see that in rural areas it may be difficult to cover the vast distances that exist between communities. The phone and internet are covering distances and spaces in ways that are physically impossible. While distance does not pose a significant barrier to care in my urban practice, consultations by phone and via e-mail are becoming more frequent. This indicates to me that such technologies assist in overcoming other types of access barriers (e.g., the availability of time), an outcome of which is that I am having to modify my practice style and particularly how I develop a trusting and potentially therapeutic relationship with patients.

I have practiced very differently in inner-city neighbourhoods as compared to semi-urban areas. The populations I have provided care for in the inner-city have required a very different style of practice in that in these areas certain social problems typically precipitate and/or exacerbate people's medical problems. Although the longitudinal relationship I tried to develop with my patients when in inner-city practice was crucial, I found that patients were more transient, frequently moved away and often did not speak English well (in that major metropolitan centres in Canada tend to attract large numbers of newcomer immigrants, many of whom are not conversant in English). My knowledge of the community became central to understanding what and why events happened to these patients. I would often draw on community resources specific to the inner-city (e.g., immigrant services, homeless shelters) to stabilize unwell individuals, or refer them to support groups or help that was available only to specific groups (e.g., those receiving social assistance). In my current semi-urban practice, the patients I encounter are more fixed geographically and tend to move around less. The longitudinal relationship that is developed with these patients seems to be a main factor in their satisfaction with the consultations we have. These patients do place different demands on my clinical skills, though. For example, patients who are more educated and of higher socio-economic standing expect a different quality of care and typically present with very different problems that are more related to disease prevention than disease treatment.

In negotiating through the many complexities of peoples' illnesses and symptoms that they present with, I have found the use of space, and spatio-temporal distance in particular, very important. At times when I am unsure of a diagnosis (and this is very common for all GPs), I use the privilege that I have of seeing the patient again at a later date to help sort out the features of the problem. In this way, the distance that I am afforded helps me to clarify my thoughts and enables me to prioritize facts and hopefully reach the correct diagnosis. In many cases, the origin of the presenting complaint is not physical or medical, but is social. My understanding of the complexities of the patient's life and family, and my ability to construct a mental map of each patient's environment enables me to reach this diagnosis and specifically to understand his/her social and life experiences.

*Implications*

The geographies of family medicine that I have outlined in this chapter and shared in the above personal reflection have implications for clinical practice. The landscape of the community can significantly change a GP's style and nature of practice, although all doctors aspire to and try to adhere to certain principles such as the four discussed above. It can also affect the numbers of GPs that practice in a particular location and the services available. Because of this, primary care training should either be targeted specifically to practitioners who expect to work in certain areas (i.e., rural versus urban) or should be very broadly taught to ensure that all family doctors have the skills they may need to practice in any location, while immersed in any community, and when seeing any population.

PHC, the first tier of health care, is diverse. Policy makers and planners must be conversant with the way that the geographies of family medicine and other spatial aspects of PHC can affect care delivery and clinical practice in multiple ways. An understanding of place can allow for tailored primary care that will suit certain locations, practitioners, and patients. GPs and other practitioners are able to act as resources for the community, and can even ration the care that they provide in a way that best meets community *and* patient needs. GPs in particular act as the gatekeepers of medicine more broadly and providers of the first level of care. Therefore, they should be a key part in decisions regarding PHC policy, because of their expert knowledge and role. In doing so they can bring to the fore issues central to the geographies of family medicine, including how they are embedded within the principles of family medicine shared throughout this chapter.

## Acknowledgements

Gina was funded by a Canadian Diabetes Association Postdoctoral Fellowship during the writing of this chapter.

## References

Aaronson, J.W., Murphy-Cullen, C.L., Chop, W.M. and Frey, R.D. (2001), 'Electronic Medical Records: The Family Practice Resident Perspective', *Family Medicine* 33:2, 128–32.

Agarwal, G. and Kaczorowski, J. (2006), 'Community Health Awareness of Diabetes (CHAD) – Feasibility and Acceptability of a Program to Improve Diabetes Risk Assessment Using Pharmacy-based, Volunteer-run Sessions', Abstract No. 87. National Young Investigators Forum, CIHR, *Exp Clin Cardiol* 11:1, 54.

Anderson, M. and Rosenberg, M.W. (1990), 'Ontario's Underserviced Area Program Revisited: An Indirect Analysis', *Soc Sci Med* 30:1, 35–44.

Arborelius, E. and Bremberg S. (1992), 'What Can Doctors do to Achieve a Successful Consultation? Videotaped Interviews Analysed by the "Consultation Map" Method', *Fam Pract* 9, 61–6.

Balint, M. (1972), *The Doctor, His Patient and the Illness* (New York: International Universities Press).

CIHI (2005), 'Geographic Distribution of Physicians in Canada. Beyond How Many and Where' (Ottawa, ON: Canadian Institute for Health Information).

Conradson, D. (2003), 'Geographies of Care: Spaces, Practices, Experiences', *Social & Cultural Geography* 4:4, 451–4.

Coyte, P.C., Catz, M. and Stricker, M. (1997), 'Distribution of Physicians in Ontario: Where are There too Few or too Many Family Physicians and General Practitioners?', *Can Fam Physician* 43, 677–83.

Cronin, A.J. (1978), *Short Stories from Dr Finlay's Casebook* (Pearson English Language Teaching).

Crooks, V.A. (2006), '"I Go on the Internet; I Always, You Know, Check to See What's New": Chronically Ill Women's Use of Online Health Information to

Shape and Inform Doctor-patient Interactions in the Space of Care Provision', *ACME: An International E-Journal for Critical Geographies* 5:1, 50–69.

Curtis P. and Rogers J. (1980), 'The Concept and Measurement of Continuity in Primary Care', *Am J of Public Health* 70, 122–7.

Euract (2005), *Eurpoean Definition of General practice/Family Medicine* (Network within WONCA Europe), 2.2 The Speciality of Family Medicine, 6.

Evans, R.G. and Stoddart, G.L. (1990), 'Producing Health, Consuming Health Care', *Soc Sci Med* 31:12, 1347–63.

Evans, R.G., Barer, M. and Marmot, T.R. (eds) (1994), *Why are Some People Healthy and Others Not? The Determinants of Health of Populations* (New York: Aldine de Gruyter).

Freeman, G. and Hjortdahl, P. (1997), 'What Future for Continuity of Care in General Practice?', *BMJ* 314:7098, 1870.

Haggerty, J.L., Reid, R.J. and Freeman, G.K. et al. (2003), 'Continuity of Care: A Multidisciplinary Review', *BMJ* 327:7425, 1219–1221.

Hansen, M.F. (1975), 'Continuity of Care in Family Practice', *J Fam Pract* 2, 439–44.

Hennen, B.K. (1975), 'Continuity of Care in Family Practice; Part 1; Dimensions of Continuity', *Journal of Family Practice* 2:5, 371–2.

Holden, H. (2004), 'The Social Geography of General Practice', *Br J Gen Pract* January, 54:498, 72–5.

Illich, I. (1976), *Medical Nemesis* (New York: Pantheon).

Kearns, R. and Moon, G. (2002), 'From Medical to Health Geography: Novelty, Place and Theory After a Decade of Change', *Prog Hum Geogr* 26, 605–27.

Krupa, L. and Chan, B. (2005), 'Canadian Rural Family Medicine Training Programs: Growth and Variation in Recruitment', *Can Fam Phys* 51, 853.

MacQueen, K., McLellan, E. and Metzger D.S. et al. (2001), 'What is Community? An Evidence-based Definition for Participatory Public Health', *American Journal of Public Health* 91:12, 1929–38.

Mainous III, A.G., Koopman, R.J. and Gill, J.M. et al. (2004), 'Relationship Between Continuity of Care and Diabetes Control: Evidence From the Third National Health and Nutrition Examination Survey', *Am J Public Health* 94:1, 66–70.

Manuel, D., Maaten, S. and Thiruchelvam, J. et al. (2006), *Primary Care in the Health Care System* (ICES Atlas), 1:11.

McWhinney, I. (1997), *A Textbook of Family Medicine*, 2nd edition (Oxford: Oxford University Press).

Mechanic, D. (2001), 'How Should Hamsters Run? Some Observations About Sufficient Patient Time in Primary Care', *BMJ* 323, 266–8.

Ogilvie, G.S., Shaw, E.A. and Lusk, S.P. et al. (2004), 'Access to Colposcopy Services for High-risk Canadian Women: Can We Do Better?', *Can J Public Health* 95:5, 346–51.

Parchman, M.L. and Burge, S. (2002), 'Continuity and Quality of Care in Type 2 Diabetes', *J Fam Pract* 51:7, 619–24.

Parr, H. (2000), 'Interpreting the Hidden Social Geographies of Mental Health: Ethnographies of Inclusion and Exclusion in Semi-institutional Places', *Health & Place* 6, 225–237.

Pemberton, J. (1969), 'William Pickles', *J R Coll Gen Pract* 18:84, 5–8.

Pereira Gray, D., Evans, P. and Sweeney, K. et al. (2003), 'Towards a Theory of Continuity of Care', *Journal of the Royal Society of Medicine* 96, 160–6.

Philo, C. (1997), 'Across the Water: Reviewing Geographical Studies of Asylums and Other Mental Health Facilities', *Health & Place* 3, 73–89.

Reid, R., Haggerty, J.L. and McKendry, R. (2002), *Defusing the Confusion: Concepts and Measures of Continuity of Healthcare* (Canadian Health Services Research Foundation).

Rosser, W. (2006), 'Sustaining the 4 Principles of Family Medicine in Canada', *Can Fam Physician* 52:10, 1191–2.

Saultz, J.W. and Lochner, J. (2005), 'Interpersonal Continuity of Care and Care Outcomes: A Critical Review', *Ann Fam Med* 3, 159–66.

Vinten-Johansen, P., Brody, P. and Paneth, N. et al. (2003), *Cholera, Chloroform, and the Science of Medicine: A Life of John Snow* (OUP).

Wall, E.M. (1981), 'Continuity of Care and Family Medicine', *J Fam Pract* 13, 655–64.

## Internet-based References

City of Hamilton, Community Health Bus (2007). Available at <http://www.myhamilton.ca/myhamilton/CityandGovernment/HealthandSocialServices/PublicHealth/CommunityHealthBus/>.

College of Family Physicians of Canada. 'Four principles of family medicine'. Section of Teachers of Family Medicine Committee on Curriculum. College of Family Physicians of Canada; 2003 [cited July 17, 2006], 8–10. Available at <http://www.cfpc.ca/English/cfpc/about%20us/principles/default.asp?s=1>.

Hippocrates. 'On airs, waters and places. 400BC', translated by Francis Adams 1796–1861. The Internet Classics Archive, HTML at Internet Classics. Available at <http://classics.mit.edu/Hippocrates/airwatpl.1.1.html>.

Chapter 8

# The Place of Nursing in Primary Health Care

Jennifer Lapum, Sandra Chen, Jessica Peterson,
Doris Leung, and Gavin J. Andrews

There is widespread inclusion of the principles of primary health care (PHC) within nursing practice. The foundational underpinnings of nurses' ways of thinking and logics of care incorporate a PHC philosophy. It is problematic to clearly understand how nurses specifically enact PHC because it is interwoven throughout their practice. The critical juxtaposition of nursing and PHC is important to consider and serves as the focus in this chapter.

An important context to the current chapter – focused on place, PHC and nursing – is a geographical perspective in nursing research. The origins of this perspective can be located in Liaschenko's early work on home care (1994; 1996; 1997). Others have provided scholarly arguments to solidify the geography of nursing as a distinct and progressing body of work (Andrews 2002; 2006; Andrews and Moon 2005; Carolan et al. 2006; Andrews and Evans 2008; Solberg and Way 2007). Not unlike human geography and the sub-discipline of medical/health geography, space and place are key concepts that underpin nursing research. While at a macro-scale space is recognized to possess mathematical and geometrical dimensions on which phenomena (such as services and diseases) can be located and distances mapped, at the micro-scale it is recognized to have significance in terms of how individuals – in this case nurses – navigate their daily practice (Halford and Leonard 2003). Places are more than locations, they are recognized as complex social phenomena with which individuals and groups develop attachments, identity and associations, and attribute meaning and significance (Cheek 2004; Gilmour 2006). Assumptions underlying PHC suggest a need to draw from a scope of knowledges, emphasizing spatial components, in order to obtain an unabridged view of care recipients. These knowledges emanate from more than just a biological basis of disease, but also an understanding of the personal biography of individuals and the social conditions that surround them (Liaschenko and Fisher 1999). A geographical perspective broadens these ways of knowing and can assist to clearly articulate the intersections of nursing and PHC.

A range of relationships between nursing and place have been articulated, including how places characterize particular nursing specialties (Montgomery 2001; Cheek 2004; Duke and Street 2003; Burges et al. 2007), impact nurse-client interactions and relationships (Purkis 1996; Liaschenko 1994; 1997; 2003; Peter 2002; Malone 2003; Peter and Liaschenko 2004), and shape interprofessional interactions and relationships (Sandelowski 2002; Brodie et al. 2005; West and

Barron 2005; Barnes and Rudge 2005). Moreover other studies have articulated the spatial embeddedness of clinical practice (Skelly et al. 2002; Affonso et al. 2004; Gesler et al. 2004; Leipert and Reutter 1998; Bender et al. 2007), while attention has also been paid to the longer-term spatial features of career paths and the cultural, social and economic forces that shape them on local (Brodie et al. 2005), national (Radcliffe 1999) and global scales (Kingma 2006).

This chapter contributes to this body of work by employing a geographical perspective and identifying issues in a so far unexplored empirical area of nursing geography and PHC. We consider how the profession of nursing has long maintained a philosophy that parallels that of PHC, but it has not necessarily been identified as such. In order to depict the affinity of nursing to a philosophy and practice of PHC, we explore the unfolding of PHC as it intersects with the historical origins of nursing and the biomedical model. Threaded throughout this chapter is our assumption that a strong presence of the biomedical model prohibits the full actualization of nurses' practice of PHC. As well, we critically examine how the philosophical and theoretical foundations of nursing are congruent with PHC. Next we move to recognize nursing as PHC, explicating how a philosophy of PHC is embedded in nurses' logics of care and practice regardless of the settings wherein they work. The chapter closes by taking a critical look forward. It is salient to preface that in our chapter, the place of nursing has two meanings. First, in a non-spatial sense, place refers to the profession's position in a particular form of health care. Second, in the spatial sense, place refers to how geographies of nursing map onto geographies of PHC.

## Situating Primary Health Care

The formalization of PHC occurred in 1978 at the International Conference of PHC (WHO 1978). The outcomes of this conference led to global initiatives and a philosophical reorientation regarding the conceptualization of health and health care systems. Prior to PHC, epistemologies that informed medicine and other health care practices (such as nursing) were grounded in biomedicine. Biomedicine is an epistemologically-informed method of practice that focuses assessment and treatment on the biological body, lending itself to objective ways of knowing. It is associated with increasingly technoscientific practice and an extension of medical jurisdiction over health and illness (Clarke et al. 2003). At times, these practices can neglect spatial forms and patterns of health particularly related to localized epistemologies. Alternatively, PHC incorporates a focus on health and illness that takes into account geographical concepts such as space and place; this does not mean that biomedical ways of knowing are obsolete. Rather, this indicates that health care practices need to incorporate multiple ways of knowing.

Because PHC and primary care are often used interchangeable in the literature, it is important to clearly articulate the distinctions and areas of overlap. Components of health care are often divided into primary, secondary and tertiary practices (Potter and Perry 1997). Primary care involves a focus on health promotion and disease prevention involving components such as immunization, sex education and flu vaccination programs, while secondary care focuses on the detection and treatment of disease and tertiary care focuses on rehabilitation and recovery (Potter and Perry 1997). Health care systems often focus attention on secondary components to the neglect of primary care. PHC re-orientates focus on primary practices, but involves secondary and tertiary components as well. More specifically, PHC is a comprehensive, philosophical approach involving principles such as: (1) health as a state of physical, mental and social wellbeing; (2) health promotion and disease prevention; (3) approaches to health care that are team-based and reflective of the sociocultural context; and (4) self-reliance and self-determination of individuals and communities (WHO 1978).

## The Unfolding of PHC: Nurses' Roles

When first introduced, PHC was an attractive idea with intentions to create healthier populations and address existing disparities. Yet, its nature proved equivocal, creating problems when translating its philosophy into practice. The profession of nursing was brought into PHC irrespective of its historical origins and foregoing place in health care systems. As such, the uptake of PHC by nurses was shaped by the early historical origins of nursing and the pre-eminent presence of the biomedical model.

*Historical Origins of Nursing*

Nursing began as a societal expectation of women's work, caring for sick family members and assisting people to give birth in the home (Berghs et al. 2006; Gordon and Nelson 2006a). In the nineteenth century, it was framed by a civic and patriotic discourse in which nursing was a public health response to attend to people deemed poor in places where social conditions were at odds with peoples' morals (Nelson 1997). At this time, nursing was already striving to address inequities and access to care as evidenced by nurses going to the places that people lived and worked (Gordon and Nelson 2006b).

Women's social status and the gendered hierarchy have strongly interfaced with the landscape of nursing (Rodney et al. 2004). The tradition of nurses as being predominantly female has continued (Rodney, Brown and Liaschenko 2004), which has led to the profession being significantly burdened with gendered expectations as they strive for equality in the workplace (Moland 2006). Secondary roles of nurses were reinforced by the initial training done by male physicians with expectations to assist obediently and unassertively (Berghs et al. 2006; Kaufman 2002). This training focused on accomplishing biomedical tasks that supported treatment and cure. The hierarchy has lessened today, but is still seen in various nurse-physician relationships. For example, nurses may make recommendations with regard to medical treatments, but physicians hold the authority to make the final decision (Moland, 2006). Nurses work within a specific scope of practice, which outlines what activities they are educated and authorized to perform (Canadian Nurses Association, 2008). However, nurses commonly indicate that even the nursing scope of practice is impinged upon by medicine (Scott et al. 2008).

Historically, it was assumed that nurses did not require education because their work was viewed as intuitive and inherently female and thus based on virtues of compassion and nurturing (Gordon and Nelson 2006a). As nurses moved beyond the handmaiden image, they strove for increased education (Gordon and Nelson 2006a). Over time the training of nurses was formalized, first within hospitals and later within academic institutions. Education contributed to the professionalization of nursing as a discipline that was self-governed with its own knowledge base and philosophical values. However, the presence of biomedicine still influences how nurses practice and their uptake of PHC.

*Presence of the Biomedical Model*

Although health care systems are different between and within countries, PHC and nursing alike are embedded in and remain strongly shaped by the biomedical model. Early origins of this model began in the seventeenth century with the Cartesian approach to the body as a biological machine in which disease was seen as a malfunction (Capra 1982). The emergence of this model gave rise to quantifiable and objective diagnostics providing an in-depth understanding of the human body. A number of benefits emanated including scientific and technological advancements of therapeutics. However, the biomedical model also led to approaches of care that lessened the place for dialogue with clients about their experiences, attitudes and expectations (Foss 2002).

There is a recognizable dissonance between a biomedical and a PHC approach. This dissonance involves a move towards a comprehensive definition of health with importance placed on health promotion and a reconfiguration of individuals as more than passive recipients of care. Within the PHC philosophy, individuals and communities are considered active participants (WHO 1978). Engaging people as active members of their health care has led many practitioners to switch from using the designation of patient to client, which is enacted in this chapter. The usage of the term patient tends to assume passivity, while client is more consistent with PHC in that individuals are a part of decisions concerning their health. The biomedical model contributed to approaches of practice that were systematic and evidence based, but it also tended to undervalue complexities and context-dependent dimensions of health (DiBartolo 1998; Lapum 2005), including social, psychological and environmental factors (Barbour 1995). Epistemologies informed by biomedicine tend to neglect the space and place of individuals. Contrarily, PHC attends to these broader geographies of health. There is a dynamic and relational nature of place and health in which the context of peoples' lives are critical to consider (Cummins et al. 2007). Thus, nurses and other practitioners must continually integrate biomedical ways of knowing with the more geographical perspective of PHC.

The implementation of PHC has been problematic as it is juxtaposed with attempts to challenge medical elitism (Cueto 2004). There is ambiguity about nurses' practice and how it is complementary to, but not the same as, medicine (Carryer et al. 1999). Nurses' practice of PHC is restrained by the historical hierarchies of biomedicine. In the United Kingdom and Australia, nurses' scope and nature of practice within a PHC philosophy is limited by their position as employees of physicians (Halcomb, Patterson and Davidson 2006). Although nurses are generally not employees of physicians in North America, they remain influenced by a biomedical approach to care. The strong tradition of the biomedical model has not been replaced by PHC, but the two have become co-present with the former holding more influence on health care practices. However, there is growing awareness across the health and social sciences that single disciplinary perspectives are inadequate due to the complex social and cultural issues implicated in current health care environments (Andrews and Evans 2008). Although a biomedical model is salient to diagnostics and therapeutics, augmenting the system with a philosophy of PHC can benefit clients, communities and practitioners.

Nurses are particularly effective at balancing and integrating their own epistemologies with biomedicine and PHC. Although nurses work within systems dominated by biomedicine, they have an affinity to a wide body of disciplinary traditions and epistemologies (Carolan, Andrews and Hodnett 2006; DiBartolo 1998). Some of these include anthropology, sociology, psychology and the humanities. Nurses have been referred to as boundary workers because they are aware of and integrate multiple epistemologies and moral visions that are often at odds with each other (Peter and Liaschenko 2004). Nursing has been shaped by humanistic-oriented epistemologies that engage the art of nursing (Johnson 1994) but also incorporate science as a foundation (DiBartolo 1998).

## The Theoretical and Philosophical Parallels Between PHC and Nursing

The theoretical and philosophical foundations of nursing closely parallel those of PHC (Besner 2004; Carryer et al. 1999). It is these close parallels that position nursing as salient to PHC. Nurses, like other practitioners, help people to maintain and regain their health. However, the values that underlie the logics and practice of nurses are distinctive from other practitioners. Specifically, the theoretical foundations of nursing have long focused attention on the nurse-client relationship (Allen 2004; Johnson 2004; Moyle, Barnard and Turner 1995). The development of these relationships provides nurses with opportunities to understand the place and space of clients and their communities. The foundations of nursing practice that are congruent with the World Health Organization's (WHO) definition of

PHC are explored herein by drawing on four concepts: (1) health; (2) agency; (3) place; and (4) access.

*Health*

The philosophy of PHC is based on a comprehensive approach to care with attention to the whole person including physical, psychological and social dimensions (Health Canada n.d.; WHO 1978). Caring for the whole person has been reported by health care users to be a key indicator of high quality PHC (Wong et al. 2008). Nursing foundations have historically incorporated broad understandings of health that are congruent with PHC. The view of a person as a whole underpins nursing theory which directs assessment at these various dimensions that influence individuals' abilities to achieve optimum health and recover from illness (Besner 2004).

Conceptualizations of health within PHC move beyond a disease-oriented model to include a focus on health promotion (WHO 1978), which is similar to the foundations of nursing. In the nineteenth century, Nightingale recognized that nursing practice involved ensuring conditions wherein health could be preserved as well as restored (Besner 2004). Current nursing theories reiterate a need for the attainment and restoration of health through the promotion of wellness, as well as treatment and rehabilitation practices (Kozier, Erb and Olivieri 1991; Potter and Perry 1997). Although PHC principles call for a shift away from illness care towards health promotion (Besner 2004), it does not negate curative health care.

Caring for the various dimensions of a person can be problematic in health care systems that are characterized by biomedical knowing. Despite the benefits of the biomedical model, it can lead to a neglect of the place and space of individuals and the impact of geography on the meanings of health. PHC maintains a different philosophy of care and requires reconceptualizations of health and how care is delivered (MacIntosh and McCormack 2000). The trend of specialization in which health care practices are divided into medical areas based on disease, a person's age or biological system is evident in nursing (Besner 2004; Carryer et al. 1999). This type of specialization parallels a biomedical model. However, nursing and PHC alike are informed by philosophies other than biomedicine (Besner 2004). Nursing is closely aligned with PHC in ways that the client is viewed as a whole and health care incorporates health promotion along with treatment of disease. The development of nurse-client relationships places nurses as critical to PHC as they are positioned to better understand the spatial components of clients' lives and their health.

*Agential Individuals and Communities*

The PHC approach promotes a system of individuals and communities who have a moral right and obligation to participate in the maintenance of their health and planning of their health care (WHO 1978; 2003). This tenet of PHC is consistent with a philosophy of self-reliance and self-determination (WHO 1978). Clients are positioned as not only the focus of care, but constructed as active members of health care teams. Fostering clients' sense of agency is an element that underpins nursing practice. Agency involves persons taking responsibility and action in their lives; though, not all clients are capable of deliberate action. The enactment of agency exists on a continuum because it is socially and contextually based on the person and partially determined by his or her illness and location in the health care system (Liaschenko and Fisher 1999).

The participation of clients in their health care, a major tenet of PHC, has not been engaged to its fullest potential in part due to the influence of the biomedical model. The omnipotent place of this model can repress subjectivity and neglect the idiosyncratic nature of human beings and human lives. If individuals are not engaged as active members in the pursuit of their health an oppressed sense of agency can be morally reinforced (Morgan 1998). The deputy director-general of WHO during the conference at Alma-Ata posited that practitioners continue to view individuals as passive recipients of health care, which curtails the potential of PHC (Tejada de Rivero 2003).

To truly effectuate clients' sense of agency, nurses draw upon relational and contextual ways of knowing (Rodney et al. 2004) and clients' personal biographies (Liaschenko and Fisher 1999). The objective ways of knowing that are associated with the biomedical approach are enhanced by epistemologies that are rooted in local context (Brown et al. 2004; DiBartolo 1998). Consequently, nurses move beyond the view of individuals as passive recipients of care. Because nursing theories draw from several disciplines and epistemologies, nurses have been particularly effective at interfacing biomedical and localized ways of knowing. This plurality of epistemologies that are dynamic and contextual underlies nursing practice.

Fostering clients' agency is salient to ethical nursing practice and parallels conceptualizations of PHC. An understanding of the place of clients and their communities can be more effectively assessed and incorporated into health care practice when practitioners focus on interpersonal communication and the development of genuine and trusting relationships. Interpersonal communication is an element of PHC that is most often associated with satisfaction (Wong et al. 2008). Although an asymmetry of power exists between nurse and client (Rodney et al. 2004), the connection developed in the relationship (Johnson 1994) and the development of trust can enhance cooperation (Mohseni and Lindstom 2007). It is through these relationships that the fostering and enactment of agential clients and communities is made possible. As a result, self-reliance and self-determination are constructed as realized dimensions of health.

Underpinnings of nursing further reflect the principles of PHC in that practice is directed at fostering clients and communities that are self-reliant (Besner 2004). Although levels of participation vary, clients typically want to be involved to some extent in their care (Thompson 2007). Clients have an understanding of their local situatedness and their capabilities for health promotion activities and treatment adherence. Emphasized in the fundamentals of nursing practice is the incorporation of clients' experiences and attitudes, which are essential to the successful implementation of health care (Kozier et al. 1991; Potter and Perry 1997). Fostering agency and engaging clients as self-reliant has potential to enact the ideology of individualized care and simultaneously connotes ethical practice. As such, care can be co-constituted based on clients and their space of living.

*Place*

Both PHC and nursing value place as an indubitable consideration for engaging in a comprehensive approach to health care. Human needs are open-ended (Peter and Liaschenko 2004) and clients and the places they live are relational and dynamic by nature (Laustsen 2006). The PHC approach recognizes the complex and constantly changing health needs of individuals and their environments. Likewise, the dynamic nature of environments is an important theoretical underpinning of nursing in that nurses engage the physical place and the social space of health and illness in their practice (Potter and Perry 1997). As well, demographic and epidemiological shifts can govern the flux in health care needs and how principles of PHC are implemented (Pottery and Perry 1997; WHO 2003).

A conscious perception of clients as inherently dynamic beings calls forth a system of practitioners who are ready to acclimatize to the individual and to the situation at any moment. Nurses are particularly well suited to the enactment of PHC because of their sustained closeness to clients and their emphasis on the nurse-client relationship. It is these principles of care that provide nurses with the opportunity to assess issues related to clients' spatial embeddedness and obtain insight into personal and biographical knowledges of health.

An understanding of clients and their context is developed through the nurse-client relationship (Laustsen 2006; Peplau 1952; Watson 1988). It is through the development of these relationships that nurses obtain a better understanding of how place shapes health. Although clients have always been the focus of health care systems, PHC renews a movement in which clients are not just isolated individuals with isolated diseases that need treatment (Romanow 2002). Clients are recognized as historical and experienced beings living in communities with a continuum of health needs. In order to develop healthier populations, this demands that the influence of broader social structures and the complexities of health are taken into consideration (Cueto 2004). Predominant in nursing theories and analogous to PHC is that clients' care moves beyond their biological status to incorporate the psychological, social and environmental dimensions of health (Besner 2004). The client cannot be merely assessed and treated in an objective manner that negates a sense of subjectivity and ignores contextual dimensions of health. The nurse-client relationship critically positions nurses to maximize dimensions of PHC that value clients as situated people. The sustained physical and/or moral proximity to clients gives nurses an understanding of the temporal nature of health needs (Peter and Liaschenko 2004) and the situated space and place of clients. Nurses tend to maintain a physical co-presence with clients that is often more sustained than other practitioners, particularly in hospital settings. Also, the focus on nurse-client relationships provides space for nurses to be able to develop moral proximity.

The closeness and moral proximity of nurses to clients has prompted attention to the amorphous nature of nursing practice. A defining principle of PHC is its responsiveness to the contextual situation, but also the situated dimension of clients' experiences and the temporal nature of health and illness (MacIntosh and McCormack 2000). As a result, practitioners need to readily respond in ways that take into consideration the dynamic needs of clients. However, the increasing specialization of practitioners has made them less flexible in responding to the dynamic and complex needs of clients (Romanow 2002). Yet, it is recognized that at least one practitioner on health care team needs to be flexible in their professional roles and expectations (Peter and Liaschenko 2004). Although nursing has also experienced this trend of specialization, the underlying philosophies and theories prepare nurses to shift their focus of care and role depending on the needs and situation of the client (Carryer et al. 1999; Kozier et al. 1991). This amorphous nature positions nurses to shift with the situation, client and community.

*Access*

A key principle of PHC is ensuring access to essential services (WHO 1978). This includes, but is not limited to physical, economic and cultural access to health care services and providers. A driving force behind the Declaration of Alma-Ata was to address health inequalities (Messias 2001). In the Declaration it was stated that health is a right by which governments should ensure that essential health care services are in place and people have access to these services (Pappas and Moss 2001). Ensuring universal access to health care services is one essential step in addressing health inequalities.

Since the Declaration, differences in life expectancy within and between nations are diminishing, but the gap in living conditions is widening and the number of people living in poverty has increased (Pappas and Moss 2001). This provides evidence that health inequalities are still a global issue. As well, there is an existing tension between the desire to give available resources to those in need and to equally distribute resources between everyone (Pappas and Moss 2001). At a micro-level, nurses encounter this challenge in their daily practices when making decisions regarding how to distribute their time and thus facilitate clients' access to care. The desire to give extra time to a client in need must be balanced with ensuring that others receive essential care (Allen 2004).

In order to understand health inequalities and issues of access, we need to consider the concept of place (Cummins et al. 2007). Because place is inseparable from clients' health, access to services impact upon how nurses practice. Inequalities with regard to access vary greatly based on the system of care, which can range from universal to private-based approaches. Embedded in nursing philosophies is the concept that all individuals deserve access to services, but achieving this is not always possible depending on a country's system. Nursing curriculum specifically addresses access to health care by teaching nurses the skills to work with culturally-diverse and difficult to access populations (Carryer et al. 1999). It is important to consider the environments wherein people live and the cultures they are a part of because of their social and relational embeddedness (Cummins et al. 2007).

PHC also entails having access to the most appropriate practitioner (MacIntosh and McCormack 2000). Nurses often assist with access by helping people navigate the health care system. This navigation role can include assisting people to access pathways for specific health care needs and determining the most appropriate practitioner (Allen 2004; Lewis 2004; MacIntosh and McCormack 2000). In PHC, any member of the health care team can be an initial point of access (MacIntosh and McCormack 2000). There are multiple portals of entry into the health care system such as community settings, the family doctor's office and acute and emergency settings. Often nurses are involved in all of these initial contact points (Potter and Perry 1997). As such, they must be ready to give and receive referrals from other practitioners (MacIntosh and McCormack 2000).

Access to a team of providers is also a key feature of PHC (Health Canada n.d.). A team approach augments comprehensive and continuous care by constructing ways that clients can move through the system with ease in the places where they live and work (Health Canada n.d.; WHO 1978). The various dimensions of health care that clients and communities require cannot be provided by one discipline or individual practitioners. Interdisciplinary teams work together to meet the needs of clients and may include nurses, physicians, physiotherapists, social workers, midwives, counsellors, dieticians, and psychologists. Significant features of nursing practice are adeptness to coordinate care (Potter and Perry 1997) and collaboration with multidisciplinary teams (Rodney et al. 2004). These nursing roles are particularly evident in hospitals and in public and community health settings because of nurses' sustained closeness as front-line care providers.

Access to PHC also involves the effective use of appropriate technology by nurses and other practitioners. Technology is opening access to health care systems in ways that were not previously possible. Structural, technological and conceptual changes within health care delivery have far reaching consequences for practice (Andrews and Evans 2008). It has become increasingly unnecessary for clients to be physically co-present with the practitioner. Nursing practice can happen from a distance in which the use of technology for therapeutic, acute, palliative, and rehabilitative care has flourished (Andrews and Evans 2008). Nurses are using technologies to initiate contact with clients, such as through telecommunications. Also, digital and electronic technologies are used to assess, monitor and treat clients. Technology has the potential to push the philosophy of PHC forward because it provides improved access to assessment, diagnostics and treatment to remote populations.

**Recognition of Nursing as PHC**

Although nurses did not have a principal role at Alma-Ata (Keighley 2006), they were identified as significant players to the successful implementation of PHC (Mahler 1985). As depicted through the examination of health, agency, place and access, the foundations of nursing closely parallel PHC. Nevertheless, it is still not common for nurses to be identified as PHC practitioners and for nurses themselves to readily acknowledge their role in providing PHC. An exception to this is the PHC nurse practitioner who has advanced education in this philosophy and way of practice.

Physicians were positioned as the leaders at Alma-Ata and pivotal to determining how PHC was adopted and enacted. As a result, nurses may not claim their practice to be PHC because their roles were determined by an authority outside of the nursing profession. Consequently, they were predisposed to secondary roles and responsibilities that were interpreted according to biomedical values and assumptions. This hierarchical pathway was a repetition of how nurses' roles were historically conceptualized. The foundations of nursing tended to be overlooked in the development of PHC. Although the profession of nursing with its own knowledge base has developed more strongly over the last few decades, the role of nursing in PHC remains ambiguous.

Problematic to PHC is that it has been limited to specific roles and settings and as a method of delivery. For example, family physicians are commonly identified as PHC practitioners because they have the ability to provide continuity of care over the long term and focus on preventive medicine. Similarly, nurse practitioners or case managers in community settings are often linked with providing PHC. In Western nations PHC is often associated with a designated interdisciplinary clinic usually physician led. The first nurse practitioner-led PHC clinic in Canada recently opened in Sudbury, Ontario (Ontario Ministry of Health and Long-Term Care 2006). Although initiatives like this set a precedent and are important to improving health care systems, they still reduce PHC to a setting and a method of delivery.

Our argument is that every nurse enacts a PHC philosophy in their daily logics of care and practice regardless of role or setting. PHC is engrained in nurses' practice because of the strong parallels between the foundations of both nursing and PHC. The tenets of PHC are employed across the continuum of care whether nurses are practicing in settings wherein they care for the acutely ill or the chronically ill or in settings where activities are directed at health promotion. Nurses are continually conducting assessments and engaging clients and their families as agential individuals and involving them in decisions about plans of care. An example of PHC that nurses typically provide includes educating individuals about a healthy lifestyle, such as nutrition and exercise, while providing a medically-based treatment such as chemotherapy or preparing a client to be discharged following heart surgery. In addition, surgical nurses mobilize clients as soon after surgery as possible to prevent complications such as pneumonia and blood clots. In mental health settings, nurses often assess environments of care to promote client safety. Although nurses in acute care may be focused on critical care treatments, they are constantly incorporating the PHC tenets of health promotion and disease prevention. A simple example includes repositioning clients in bed every two hours to promote blood circulation and prevent skin breakdown. As well, nurses are often critical players to help individuals navigate health care systems at any stage of care.

## A Critical Look Forward

Health care providers practise in systems that value and benefit from biomedical epistemologies. However, a focus on biomedical epistemologies can constrain how nurses practise. This manifests in something as basic as the daily documentation that nurses complete on clients. Nurses' documentation is biomedically oriented and as such, evidence of PHC activities are neglected (Tornvall, Wahren and Wilhelmsson 2007). Nursing documentation often includes objective parameters such as clients' vital signs, pain-ratings and biomedical tasks such as completion of dressing changes. The integration of PHC into biomedical-dominated systems has been problematic. Epistemologies informed by nursing and geography that emphasize health as relational and spatial have been undervalued. As well, nurses have neglected to fully recognize and identify their own practice as PHC. A PHC focus is evident in some schools of nursing. However to better serve PHC as a philosophy, we suggest nursing curricula weave principles and applications of PHC throughout, rather than relegate it to discrete courses. Only with repetition and consistency can we provide clear reinforcement and understanding of a PHC approach and how it can be integrated and complement biomedicine.

The nexus between nursing practice and PHC is not a new phenomenon. The theoretical and philosophical foundations of nursing practice have paralleled those of PHC since its inception. Without identifying it as such, nurses have long enacted principles of PHC by supporting a comprehensive understanding of health and place, fostering agential clients and communities, and employing various technological methods to enhance access to care and health care delivery methods. PHC is embedded as a philosophy of care in nursing and is not necessarily associated with a role or setting. This has positioned nurses as salient to PHC, but has also created ambiguities. Because of the ambiguities concerning PHC and nursing, empirical work is required to explore nurses' conceptualization of PHC and how it intersects with their practice. Meanings will likely be inconsistent amongst nurses internationally, reflecting nuanced understandings, spatial relationships and the various types of health care systems. As well, the development of an evaluation tool will assist in teasing apart the embedded principles of PHC in nursing practice that have long been unrecognized. Nurses need to identify themselves as practitioners who provide care consistent with PHC, particularly to legitimize and develop appropriate social structures to support their role.

PHC cannot be reduced to a method of care, but needs to be acknowledged as something practitioners engage in their logics of care and enact in their social health care practices. The pluralism of philosophies and theories that inform nursing practice provide opportunities to enact broad understandings of health, including geographically-informed epistemologies. The emphasis on moral and physical proximity and relationships with clients and communities position nurses as geographically flexible and spatially informed. They are able to draw on concepts of place and space, so that meanings of health and health care are tailored to individuals. Because of the close parallels between the theoretical and philosophical foundations of nursing and PHC, nurses are continually enacting a PHC approach in virtually every setting and role. Our argument in this chapter renders PHC as being everywhere in nurses' work, which can make it difficult to distinguish how it is specifically enacted. However, a philosophy that underlies health care systems, such as PHC, needs to be woven throughout all dimensions of social health care practices in order to be truly actualized.

It is critical that awareness is raised with practitioners and policy makers regarding how nursing practice and PHC intersect and requires skilled clinicians to draw from and integrate a plurality of epistemologies. The next step is for nurses to engage in global-level decision making and to begin informing PHC policy. With nurses' historical and on-going PHC approach to their work, it is critical that they are involved in current implementations, research, and evaluations of PHC. Not only medicine but nursing and other disciplines need to take a role in broadening the current interpretations and ownership of a PHC-oriented health care system.

Attention to PHC has occurred in waves, but there is renewed interest as the thirty year anniversary of the Declaration is upon us (Haines, Horton and Bhutta 2007). Nurses and other practitioners need to articulate how their positions can best be configured. Taking up a geographical perspective in which we critically consider the intersections between nurses, place and PHC may move us forward.

## References

Affonso, D., Andrews, G. and Jeffs, L. (2004), 'The Urban Geography of SARS: Paradoxes and Dilemmas in Toronto's Healthcare', *Journal of Advanced Nursing* 45:6, 1–11.

Allen, D. (2004), 'Re-reading Nursing and Re-writing Practice: Towards an Empirically Based Reformulation of the Nursing Mandate', *Nursing Inquiry* 11:4, 271–83.

Andrews, G. (2002), 'Towards a More Place-sensitive Nursing Research: An Invitation to Medical and Health Geography', *Nursing Inquiry* 9:4, 221–38.

Andrews, G. (2006), 'Geographies of Health in Nursing', *Health & Place* 12:1, 110–18.

Andrews, G. and Evans, J. (forthcoming), 'Understanding the Reproduction of Health Care: Towards Geographies in Health Care Work', *Progress in Human Geography*.

Andrews, G. and Moon, G. (2005), 'Space, Place and the Evidence Base, Part Two: Rereading Nursing Environment Through Geographical Research', *Worldviews on Evidence-based Nursing* 2:3, 142–56.

Barbour, A. (1995), *Caring for Patients: A Critique of the Medical Model* (Stanford, California: Stanford University Press).

Barnes, L. and Rudge, T. (2005), 'Virtual Reality or Real Virtuality: The Space of Flows and Nursing Practice', *Nursing Inquiry* 12:4, 306–15.

Bender, A., Clune, L. and Guruge, S. (2007), 'Considering Place in Community Health Nursing', *Canadian Journal of Nursing Research* 39:3, 20–35.

Berghs, M., Dierckx de Casterle, B. and Gastmans, C. (2006), 'Nursing, Obedience, and Complicity with Eugenics: A Contextual Interpretation of Nursing Mortality at the Turn of the Twentieth Century', *Journal of Medical Ethics* 32:2, 117–22.

Besner, J. (2004), 'Nurses' Role in Advancing Primary Health Care: A Call to Action', *Primary Health Care Research and Development* 5, 351–358.

Brodie, D. et al. (2005), 'Working in London Hospitals: Perceptions of Place in Nursing Students' Employment Considerations', *Social Science & Medicine* 61, 1867–81.

Brown, H., Rodney, P. and Pauly, B. et al. (2004), 'Working within the Landscape: Nursing Ethics', in J. Storch, P. Rodney and R. Starzomski (eds), *Toward a Moral Horizon: Nursing Ethics for Leadership and Practice* (Toronto: Pearson Education Canada Inc.), 126–53.

Burges-Watson, D., Murtagh, M.J. and Lally, J.E. et al. (2007), 'Flexible Therapeutic Landscapes of Labour and the Place of Pain Relief', *Health & Place* 13, 865–76.

Capra, F. (1982), *The Turning Point: Science, Society, and the Rising Culture* (New York: Simon & Schuster).

Carolan, M., Andrews, G. and Hodnett, E. (2006), 'Writing Place: A Comparison of Nursing Research and Health Geography', *Nursing Inquiry* 13:3, 203–19.

Cheek, J. (2004), 'Older People and Acute Care: A Matter of Place', *Illness, Crisis and Loss* 12:1, 52–62.

Clarke, A.E., Shim, J.K. and Mamo, L. et al. (2003), 'Biomedicalization: Technoscientific Transformations of Health, Illness, and U.S. Biomedicine', *American Sociological Review* 68:2, 161–94.

Cueto, M. (2004), 'The Origins of Primary Health Care and Selective Primary Health Care', *American Journal of Public Health* 94:11, 1864–74.

Cummins, S. et al. (2007), 'Understanding and Representing "Place" in Health Research: A Relational Approach', *Social Science & Medicine* 65:9, 1825–38.

DiBartolo, M. (1998), 'Philosophy of Science in Doctoral Nursing Education Revisited', *Journal Professional Nursing* 14:6, 350–60.

Duke, M. and Street, A. (2003), 'Hospital in the Home: Constructions of the Nursing Role: A Literature Review', *Journal of Clinical Nursing* 12, 852–9.

Foss, L. (2002), *The End of Modern Medicine: Biomedical Science Under a Microscope* (New York: State University of New York Press).

Gesler, W.M., Hayes, M. and Arcury, T.A. et al. (2004), 'Use of Mapping Technology in Health Intervention Research', *Nursing Outlook* 52:3, 142–6.

Gilmour, J. (2006), 'Hybrid Space: Constituting the Hospital as a Home for Patients', *Nursing Inquiry* 13:1, 16–22.

Gordon, S. and Nelson, S. (2006a), 'Moving Beyond the Virtue Script in Nursing', in S. Nelson and S. Gordon (eds), *The Complexities of Care: Nursing Reconsidered* (Ithaca: Cornell University Press), 13–29.

Gordon, S. and Nelson, S. (2006b), 'Nurses Wanted: Sentimental Men and Women Need Not Apply', in S. Nelson and S. Gordon (eds), *The Complexities of Care: Nursing Reconsidered* (Ithaca: Cornell University Press), 185–190.

Haines, A., Horton, R. and Bhutta, Z. (2007), '"Primary Health Care Comes of Age," Looking Forward to the 30th Anniversary of Alma-Ata: Call for Papers', *The Lancet* 370:9591, 911–3.

Halcomb, E., Patterson, E. and Davidson, P. (2006), 'Evolution of Practice Nursing in Australia', *Journal of Advanced Nursing* 55:3, 376–90.

Halford, S. and Leonard, P. (2003), 'Space and Place in the Construction and Performance of Gendered Nursing Identities', *Journal of Advanced Nursing* 42:2, 201–8.

Johnson, J. (1994), 'A Dialectical Examination of Nursing Art', *Advances in Nursing Science* 17:1, 1–14.

Johnson, J. (2004), 'Philosophical Contributions to Nursing Ethics', in J. Storch, P. Rodney and R. Starzomski (eds), *Toward a Moral Horizon: Nursing Ethics for Leadership and Practice* (Toronto: Pearson Education Canada Inc.), 42–55.

Kaufman, G. (2002), 'Investigating the Nursing Contribution to Commissioning in PHC', *Journal of Nursing Management* 10:2, 83–94.

Keighley, T. (2006), 'From Sickness to Health', in S. Nelson and S. Gordon (eds), *The Complexities of Care: Nursing Reconsidered* (Ithaca: Cornell University Press), 88–103.

Kingma, M. (2006), *Nurses on the Move: Migration and the Global Health Care Economy* (Ithaca: Cornell University Press).

Kozier, B., Erb, G. and Olivieri, R. (1991), *Fundamentals of Nursing: Concepts of Process and Practice* (Redwood City, California: Addison-Wesley).

Lapum, J. (2005), 'Women's Experiences of Heart Surgery Recovery: A Poetical Dissemination', *Canadian Journal of Cardiovascular Nursing* 15:3, 12–20.

Laustsen, G. (2006), 'Environment, Ecosystems, and Ecological Behavior: A Dialogue Toward Developing Nursing Ecological Theory', *Advances in Nursing Science* 29:1, 43–54.

Liaschenko, J. (1994), 'The Moral Geography of Home Care', *Advances in Nursing Science* 17:2, 16–25.

Liaschenko, J. (1996), 'A Sense of Place for Patients: Living and Dying', *Home Care Provider* 1:5, 270–2.

Liaschenko, J. (1997), 'Ethics and the Geography of the Nurse-patient Relationship: Spatial Vulnerable and Gendered Space', *Scholarly Inquiry for Nursing Practice* 11:1, 45–59.

Liaschenko, J. (2003), 'At Home with Illness: Moral Understandings and Moral Geographies', *Ethica* 15:2, 71–82.

Liaschenko, J. and Fisher, A. (1999), 'Theorizing the Knowledge that Nurses Use in the Conduct of their Work', *Scholarly Inquiry for Nursing Practice: An International Journal* 13:1, 29–41.

Leipert, B. and Reutter, L. (1998), 'Women's Health and Community Health Nursing Practice in Geographically Isolated Settings: A Canadian Perspective', *Health Care for Women International* 19, 575–88.

MacIntosh, J. and McCormack, D. (2000), 'An Integrative Review Illuminates Curricular Applications of Primary Health Care', *Journal of Nursing Education* 39:3, 116–23.

Mahler, H. (1985), 'Nurses Lead the Way', *World Health* 28:2.

Malone, R. (2003), 'Distal Nursing', *Social Science & Medicine* 56:11, 2317–26.

Messias, D. (2001), 'Globalization, Nursing, and Health for All', *Journal of Nursing Scholarship* 33:1, 9–11.

Mohseni, M. and Lindstom, M. (2007), 'Social Capital, Trust in the Health-care System and Self-rated Health: The Role of Access to Health Care in a Population-based Study', *Social Science & Medicine* 64:7, 1373–83.

Moland, L. (2006), 'Moral Integrity and Regret in Nursing', in S. Nelson and S. Gordon (eds), *The Complexities of Care: Nursing Reconsidered* (Ithaca: Cornell University Press), 50–68.

Montgomery, P. (2001), 'Shifting Meaning of Asylum', *Journal of Advanced Nursing* 33:4, 425–31.

Morgan, K. (1998), 'Contested Bodies, Contested Knowledges: Women, Health, and the Politics of Medicalization', in S. Sherwin (ed.), *The Politics of Women's Health: Exploring Agency and Autonomy* (Philadelphia: Temple University Press), 83–121.

Moyle, W., Barnard, A. and Turner, C. (1995), 'The Humanities and Nursing: Using Popular Literature as a Means of Understanding Human Experience', *Journal of Advanced Nursing* 21, 960–4.

Nelson, S. (1997), 'Pastoral Care and Moral Government: Early 19th Century Nursing and Solutions to the Irish Question', *Journal of Advanced Nursing* 26:1, 6–14.

Pappas, G. and Moss, N. (2001), 'Health for all in the Twenty-first Century, World Health Organization Renewal, and Equity in Health: A Commentary', *International Journal of Health Services* 31:3, 647–58.

Peplau, H. (1952), *Interpersonal Relations in Nursing* (New York: GP Putnam's Sons).

Peter, E. (2002), 'The History of Nursing in the Home: Revealing the Significance of Place in the Expression of Moral Agency', *Nursing Inquiry* 9:2, 65–72.

Peter, E. and Liaschenko, J. (2004), 'Perils of Proximity: A Spatiotemporal Analysis of Moral Distress and Moral Ambiguity', *Nursing Inquiry* 11:4, 218–25.

Potter, P. and Perry, A. (1997), *Fundamentals of Nursing: Concepts, Process, and Practice*, 4th edition (St. Louis: Mosby-Year Book).

Purkis, M. (1996), 'Nursing in Quality Space: Technologies Governing Experiences of Care', *Nursing Inquiry* 3, 101–11.

Radcliffe, P. (1999), 'Geographical Mobility, Children and Career Progress in British Professional Nursing', *Journal of Advanced Nursing* 30:3, 758–68.

Rodney, P., Brown, H. and Liaschenko, J. (2004), 'Moral Agency: Relational Connections and Trust', in J. Storch, P. Rodney and R. Starzomski (eds), *Toward a Moral Horizon: Nursing Ethics for Leadership and Practice* (Toronto: Pearson Education Canada Inc.), 154–77.

Romanow, R. (2002), *Building on Values: The Future of Health Care in Canada*. Commission on the Future of Health Care in Canada (ed.) (Final Report, National Library of Canada), 356.

Sandelowski, M. (2002), 'Visible Humans, Vanishing Bodies and Virtual Nursing: Complications of Life, Presence, Place and Identity', *Advances in Nursing Science* 24:3, 58–70.

Scott, S.D., Estabrooks, C.A. and Allen, M. et al. (2008), 'A Context of Uncertainty: How Context Shapes Nurses' Research Utilization Behaviors', *Qualitative Health Research* 18:3, 347–57.

Skelly, A.H., Arcury, T.A. and Gesler, W.M. et al. (2002), 'Sociospatial Knowledge Networks: Appraising Community as Place', *Research in Nursing and Health* 25, 159–70.

Solberg, S. and Way, C. (2007), 'The Importance of Geography and Health in Nursing Research', *Canadian Journal of Nursing Research* 39:3, 13–8.

Tejada de Rivero, D. (2003), 'Alma-Ata Revisited', *Perspectives in Health Magazine* 8:2, 1–7.

Thompson, A. (2007), 'The Meaning of Patient Involvement and Participation in Health Care Consultations: A Taxonomy', *Social Science & Medicine* 64, 1297–1310.

Tornvall, E., Wahren, L. and Wilhelmsson, S. (2007), 'Impact of Primary Care Management of Nursing Documentation', *Journal of Nursing Management* 15:6, 634–42.

Watson, J. (1988), *Nursing: Human Science and Human Care: A Theory of Nursing* (New York: National League for Nursing).

West, E. and Barron, D. (2005), 'Social and Geographical Boundaries around Senior Nurse and Physician Leaders: An Application of Social Network Analysis', *Canadian Journal of Nursing Research* 37:3, 132–48.

Wong, S., Watson, D., Young, E. and Regan, S. (2008), 'What Do People Think is Important about Primary Healthcare?', *Healthcare Policy/Politiques de Sante* 3:3, 89–104.

**Internet-based References**

Canadian Nurses Association (2008), *The Practice of Nursing.* Retrieved on February 15, 2008 from http://www.cna-nurses.ca/CNA/practice/scope/default_e.aspx.

Carryer, J. et al. (1999), *Locating Nursing in Primary Health Care: A Report for the National Health Committee.* Retrieved on May 10, 2007 from http:/www.nhc.govt.nz/Publications/phc/phcnursing.pdf.

Health Canada (n.d.), About Primary Health Care (p. 3). Retrieved on February 4, 2007 from http://www.hc-sc.gc.ca/hcs-sss/prim/about-apropos/index_e.html#1.

Lewis, S. (2004), *A Thousand Points of Light? Moving Forward on Primary Health Care.* Paper presented at the National Primary Health Care Conference, Winnipeg, Manitoba. Retrieved on May 12, 2007 from www.manitobapha.ca/Final_Synthesis_Report_eng_oct13.pdf.

Ontario Ministry of Health and Long-Term Care (2006), McGuinty Government Improving Access to Primary Health Care in Northeastern Ontario. Retrieved on August 10, 2007 from http://www.health.gov.on.ca/english/media/news_releases/ archives/nr_06/nov/nr_111006.html.

World Health Organization (1978), *Declaration of Alma-Ata: International Conference on Primary Health Care, Alma-Ata, USSR, 6–12 September.* Retrieved January 10, 2007 from www.who.int/hpr/NPH/docs/declaration_almata.pdf.

World Health Organization (2003), *The World Health Report 2003: Shaping the Future.* Geneva: World Health Organization. Retrieved January 10, 2007 from http://www.who.int/whr/2003/en/whr03_en.pdf.

# Chapter 9

# Reinventing Primary Care:
# The New Zealand Case Compared

Ross Barnett and Pauline Barnett

In 2004 the Commonwealth Fund's annual international health system survey continued its focus on public satisfaction with the availability and quality of health care in five western countries (Schoen 2004). These countries, including New Zealand, have differing primary care systems, ranging from those that rely on free market models to those characterized by more government regulated systems. Of particular interest in this survey was the sharp increase in public satisfaction with the provision of health care in New Zealand, a gain that was unmatched by any of the other surveyed countries (the United States [US], United Kingdom [UK], Canada and Australia). While many factors influence public satisfaction with health services, it was no coincidence that this change followed a period of major restructuring in the funding and provision of primary care. Particularly important has been the development of the New Zealand Primary Health Care Strategy (King 2001). This signalled a move away from an individualized fee-for-service model to a more community health approach in which general practitioners (GPs), the main primary care providers, have become less dominant decision-makers, for example, in setting fees. The New Zealand Primary Health Care Strategy could not be characterized as a fully comprehensive primary health care approach, consistent with all Alma-Ata Declaration principles articulated by the World Health Organization (WHO 1978), but it nevertheless represents a movement beyond a narrow general practice model towards a broader conceptualization of primary care. This has been an important development, aimed at reorientating patterns of service delivery and producing more equitable access to primary care, particularly for vulnerable populations.

These changes represent an important milestone in the evolution of primary care in New Zealand and raise questions regarding the impact of such reforms not only upon GP and other primary care providers but also the populations they serve. Given this, it is the purpose of this chapter to review the development of primary care in New Zealand and consider this in an international context. The chapter has three specific aims:

1.  To discuss the development of primary care in New Zealand, and its recent restructuring;
2.  To draw comparisons between New Zealand and other selected countries; and

3. To discuss the extent to which the underlying factors that have shaped primary care reform in New Zealand are consistent with international experiences.

## Changes in Primary Health Care in New Zealand

This section first discusses recent changes in the organization and funding of primary care in New Zealand, including the transition from independent GPs to the formation of professional group collectives and their replacement by an alternative third sector model. These organizational changes have been accompanied by changes in funding, with less emphasis on fee-for-service and moves towards capitation and an increased focus on population health. We discuss these developments and their implications, but first examine historical barriers to good primary care in New Zealand and the conflicted nature of state attempts to address wider issues of social and ethnic inequality.

### Growth and Impact of Fee-for-Service

In the 1930s the first Labour government in New Zealand sought to create a universal welfare system in which all doctors would be paid servants of the state. GPs strongly resisted this and, after a long battle the government acquiesced and, in 1941, agreed to an alternative funding model. This involved a fee-for-service subsidy arrangement that reimbursed doctors for the cost of the GP consultation fee and funded the cost of associated pharmaceuticals and laboratory tests. These costs were demand-driven and largely uncontrolled, and practitioners had no obligations to the funder other than to see patients. They also retained the right to charge an additional fee. Hospitals and their doctor employees, however, were fully funded by the state via an annual, negotiated, allocation.

These contrasting allocation methods persisted for the next six decades, leading to problems of service fragmentation, cost shifting and both regional and local inequalities in access. Successive governments, reluctant to commit to further open-ended funding and non-accountable providers in primary care, failed to increase the fee-for-service subsidy to keep pace with the growing costs of care; by the mid-1980s only one third of the cost of a consultation was met by the government, with patient co-payments meeting the rest. This created barriers to access, lack of focus on prevention and early intervention, and inequalities in distribution and provision of GP services (Barnett 1993).

### Barriers to Effective Primary Care

Until the 1970s economic inequalities in access to primary care were masked by the full employment and economic prosperity experienced throughout New Zealand, permitting significantly higher levels of GP use by those on lower incomes (Davis 1985; 1986). However, this situation changed and by the mid-1980s there were clear class and ethnic disparities in access, with poor, Maori (i.e., New Zealand's indigenous peoples) and Pacific (i.e., immigrants from the surrounding Pacific Islands) peoples most disadvantaged (Barnett and Kearns 1996; Gribben 1992) and expressing high levels of dissatisfaction with GP fees (Barnett and Coyle 2000; Gribben 1993). Despite some targeting of subsidies from 1991 onward, there was evidence of continued unmet need among less affluent groups (Barnett and Coyle 2000; Barnett 2001). This was largely a result of the effects of economic and welfare restructuring which disproportionately affected the poor during the 1990s. Policies such as the privatization of state owned enterprises and the implementation of market rents in the social housing sector resulted in an increase in poverty and economic hardship, one of the consequences of which was to reduce the use of health services. For example, patients with financial difficulties adopted a variety of coping strategies which included a high rate of switching GPs because of cost, deferral of visits, non-collection of medication, and a lower rate of visiting among those with poorer self-rated health (Barnett 2001). There was also evidence of geographical variation in the ways some GP practices attempted to aid deprived patients. Surgeries located in more deprived areas were less flexible in the financial options (such as deferred payment schemes) they made available to patients, although they were more likely to charge lower fees. This, however, did not prevent high rates of patient mobility or significantly reduce the level of unmet need. These findings are consistent with Donelon et al. (1999) who reported New Zealand second only to the US among the countries studied with respect to financial barriers to primary care.

However, poor access could not be attributed to economic factors alone. Even in health centres providing services to Maori and low income people, utilization rates in 1994/95 were substantially lower than for the country as a whole (Malcolm 1996). A later study of low fee, capitated practices serving low income populations also reported low utilization rates, but higher for Maori than others, reflecting the tribal base of several providers (Crampton et al. 2001). Besides cultural factors, shortages of doctors in poorer areas, which place geographical barriers to access, have been well documented (Barnett 1993). Effective primary care was also inhibited by organizational and practice features. Although GPs, from the 1970s onward, increasingly worked in group practices, this did not necessarily ensure a team approach to care. The need for the doctor to see a patient prevented effective use of nurses and the fee-for-service system created a competitive and episodic approach to care, limiting preventive opportunities. The standards-setting and educational role of the Royal New Zealand College of General Practitioners was important, but the benefits of such initiatives were limited by the individual nature of general practice. This individual approach also inhibited co-ordination with the hospital system, the implications of which among disadvantaged populations being seen in the increasing social polarization of acute and re-admissions and avoidable admissions during the 1990s (Barnett and Lauer 2003), all of which might be ameliorated by improved primary care services.

*Strategies for Change*

Since the 1980s efforts to improve the performance of primary care have depended on the overall philosophy and ideology of the government of the time. Since the battle over the passage of the Social Security Act of 1938, GPs have been particularly mistrustful of Labour governments. Three phases of change can be identified: a reforming Labour government (1984–90); market-led reforms under a National (conservative) government (1990–99); and primary care restructuring under Labour since 1999.

During the period 1984–1990 the Labour government attempted to address primary care, largely through managerialist and regulatory approaches. GPs were made subject to the Commerce Act of 1986 which was intended to promote competition and reduce fees. The strategy failed to achieve this, but did increase choice and widen the supply of services through the formation of new clinics by entrepreneurial doctors (Kearns and Barnett 1997). The Government was successful in targeting the weakest point of the general practice workforce, foreign medical graduates, by requiring them to work initially in rural or disadvantaged communities, and successfully challenged the dominance of medicine in maternity services by legislating to permit midwives to practice independently in the community. Other regulatory efforts were less successful, with doctors using the courts to challenge the Government's attempt to limit doctors' fees in exchange for higher subsidies. Few doctors, in rural or more deprived areas, took up the offer of contracts in exchange for higher payments (Matheson and Hoskins 1992). Intentions to include primary care within the new structure of the health system (area health boards) were not pursued because of GP opposition.

However, some non-regulatory approaches had long term implications. In an effort to meet community needs and to provide an alternative to the so-called business model of general practice, a number of new services were funded. These included both nurse-led services and community-led, not-for-profit general practices in high need urban and rural areas, including services run for and by Maori. These third sector services were effective in meeting the needs of disadvantaged groups (Crampton et al. 2004); they challenged both the autonomy and dominance of GPs and were to have a far-reaching influence in subsequent decades. Despite this, the overall effect of government intervention between 1984 and 1990 on primary care was limited, with GPs able to retain their economic and professional autonomy and with little impact on inequalities overall.

The period of 1991 to 1999 saw radical change in the organization of primary care, with the formation of doctor collectives, independent practitioner associations (IPAs), in response to market-style health reforms. The new National (conservative) government signaled its intention to move general practice from a re-imbursement model to a contract relationship with regional funders. This created additional opportunities for third sector services which were able to extend their services with more comprehensive team approaches, but was a major challenge to the dominant business model of general practice. The response of GPs was to organize themselves into formal organizations, usually charitable companies, in order to negotiate with funders from a position of greater strength. In this they were successful, with some IPAs taking on budget-holding roles for pharmaceutical and laboratory services, and most engaging in the clinical upskilling of their members, the development of quality frameworks and the provision of management support. However, more ambitious plans to develop integrated care organizations, involving both primary and secondary care along the lines of UK fundholders or US health maintenance organizatons (HMOs), were not implemented mainly because of the Minister of Health's view that IPAs did not have sufficient business skills to assume these added responsibilities. This was confirmed by research that showed greatest financial benefit tended to occur amongst a small group of GPs who were already the most efficient managers (Malcolm 1998a).

Many GPs were reluctant to join IPAs, fearing loss of clinical and economic autonomy, but seeing little alternative. These fears were largely unrealized, with practitioners reporting that IPAs had strengthened their position in the local health system, but with fee-for-service and the right to charge fees uncompromised (Barnett 2003). Nevertheless, the individual nature of general practice had undergone some change, with the IPAs introducing a new, more managerialist dimension to primary care. While efficiency and quality improved, equity issues went largely unaddressed (Malcolm 1998b), except in the small third sector.

The third phase of reform has been under a Labour government, which returned to power in 1999. The 1984 government had been notably unsuccessful in its efforts to re-orientate general practice by tinkering with the regulatory framework. A more determined effort was made in 2001 with an overall restructuring of health services into 21 district health boards (DHBs) and the issuing of the New Zealand Primary Health Care Strategy. The Strategy provided for the formation of primary health organizations (PHOs), set up with community governance, to be the contracting agency on behalf of DHBs for general practice and other primary care services. PHOs were required to have enrolled populations, be funded on a capitation basis, and be responsible for identifying and addressing health inequalities.

The capitation formula initially provided for additional payments to PHOs serving areas of high need, or with high proportions of Maori and Pacific people in their client populations. This so-called access or high needs formula was aimed at practices in the most deprived areas. To be eligible for these higher capitation payments, PHOs either needed half of their enrollees to be Maori and/or Pacific people or half of the practice population to reside in highly deprived localities. The access formula provided a strong incentive in those areas for the early development (i.e., in 2002–03) of PHOs, many of which were based on existing third sector services. PHOs in other areas, reliant on lower capitation payments, known as the interim formula, were slower to develop, but the government's plan to provide equivalent higher payments progressively to all PHOs saw PHO formation gather pace from 2004 as significantly higher payments were made available to patients in interim PHOs (see Table 9.1). By 2006, 98 percent of the population was enrolled in PHOs. Additional funding streams were made available to improve access to high need populations (see Table 9.1) and to introduce a stronger health promotion approach into primary care.

**Table 9.1  Roll out of primary health care funding**

| Date | Extension of PHC funding |
| --- | --- |
| October 2003 | when enrollees aged between 6 and 17 years became eligible for subsidies to lower the cost of doctor visits |
| April 1, 2004 | funding for low cost pharmaceuticals for enrollees in Access-funded PHOs, and 6–17 year olds enrolled in Interim-funded PHOs (maximum charge of NZ$3 per item on subsidised pharmaceuticals) |
| July 1, 2004 | funding to lower the cost of doctor visits and pharmaceutical charges for people aged 65 years and over enrolled in Interim-funded PHOs |
| July 1, 2005 | funding to lower the cost of doctor visits and pharmaceutical charges for people aged 18–24 years enrolled in Interim-funded PHOs |
| July 1, 2006 | funding to lower the cost of doctor visits and pharmaceutical charges for people aged 45–64 years enrolled in Interim-funded PHOs |
| July 1, 2007 | funding to lower the cost of doctors visits and pharmaceutical charges for people age 25–44 years enrolled in Interim funded PHOs |

*Source*: Ministry of Health, Wellington (http://www.moh.govt.nz/moh.nsf/indexmh/phcs-funding).

This policy, with its focus on reducing inequalities and community governance, reflected the third sector rather than the dominant IPA model of primary care. However, the IPA model was critical to the implementation of the new arrangements. First, without even the short history of organized general practice in IPAs, GPs would have been more vociferous and politicized and more difficult to manage for the government. Second, if there had been no organizations already in place, the necessary support for patient enrolment and funding management would have been difficult to achieve.

The new model is based on structural reform, incorporating a strong managerialist ethos, placing doctors in a contractual relationship with PHOs with formal performance and reporting requirements. As the government increased resources to support access to primary care, it progressively limited the freedom of GPs to set fee levels, arguing that subsidies should be passed on to patients and not swallowed up by practice costs or GP salaries. Beginning in early 2005, PHOs reluctantly agreed to abide by a new fees review framework in exchange for the Government's subsidies for patients (Cameron 2007a). There is also a structured performance management programme (PMP) (District Health Boards of New Zealand 2007) which assesses the extent to which PHOs are reaching certain specified clinical, process and financial objectives. Particularly important are the clinical indicators, such as immunization levels for children and the elderly, and breast screening or cervical smears. These are high needs indicators and focus on national priority initiatives relating to particular areas of health inequality. The PMP provides limited financial rewards for performance, with payments not made directly to practitioners but to the PHOs which then decide how to distribute funds, including passing performance payments to individual practices or practitioners.

Clearly for GPs there has been some loss of both clinical and economic autonomy. In addition, changing professional relationships reflect threats to professional dominance. Building on the earlier establishment of alternative services from 1984 to 1990, the Primary Health Care Strategy explicitly promotes a more independent role for nurses, and this is reflected in particular specific funding policies and in capitation itself. Challenges to both autonomy and dominance have been reinforced by the community governance model, which promotes involvement by non-general practice providers and community interests, including Maori and Pacific people, in decision-making. This has allowed a more explicit approach to addressing inequalities, with local flexibility to target programmes to high need areas and groups; a process in which IPAs were largely not engaged.

*Challenges in the PHO Approach*

While the development of PHOs has shifted the emphasis of primary care from individuals to population health, their development presents a number of challenges. First, many PHOs are small in size (see Figure 9.1) and lack the resources to achieve effective health promotion and a reduction in health inequalities. As a consequence there is increasing co-operation among groups of small PHOs, and between small and larger PHOs, in order to ensure effective information systems, share expertise, and monitor ongoing changes in access and health outcomes.

Second, despite controls on doctors' fees and increased rates of utilization, fees remain a barrier to segments of the population accessing care, particularly the poor and the elderly. The implementation of a capitation formula did not remove co-payments but merely provided greater targeting of government resources to high needs practices. As a consequence, patient part-charges remain and even for groups such as children, nationwide only 67 percent of practices offered free care in 2007, compared to 76 percent in 2005 (Consumers' Institute 2007). Nevertheless the development of PHOs and increased patient subsidies has resulted in lower doctor's fees for many patients. In 2005 fees were lower overall and particularly so for those in more deprived communities (Gribben and Cumming 2007). As a consequence, consultation rates have increased, particularly for older people and lower socio-economic groups. The one exception has been for children under six years of age, where the increase in visits has been greater for children from better off groups.

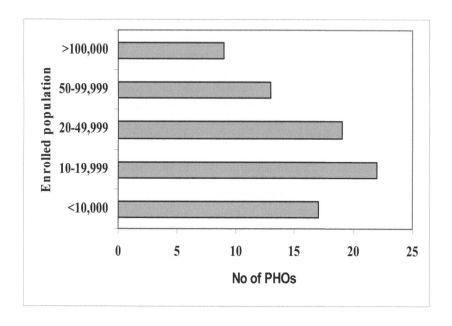

**Figure 9.1  PHOs by enrolled population size, 2006**

Third, PHOs represent a fundamental part of the New Zealand Primary Health Care Strategy, the overall goal of which is to improve health and reduce inequalities. However, there are limits to the extent that health inequalities can be reduced by changes within the health system alone. Thus, one of the key challenges for PHOs is to build wider collaborative links with other government and voluntary agencies. This is important in areas such as health promotion where lifestyle issues such as smoking prevalence and rates of obesity, for example, are affected by local environmental influences such as levels of community inequality, neighbourhood incivilities or the structure of the built environment (Macintyre et al. 2002).

Finally, unlike IPAs, PHOs are bureaucratic organizations established by government with community input designed to achieve equity objectives. A challenge is to maintain enthusiasm among GPs and other PHO members working towards these objectives. Already there is disquiet in the GP community over reporting requirements and fee control, and this is unlikely to subside. Multi-disciplinary governance also poses problems where doctors are increasingly forced to work alongside lay members on PHO boards. While IPAs and PHOs continue to coexist, relationships are sometimes uneasy. Both retain contracts with DHBs but IPAs, once the dominant GP-controlled provider organization, are now increasingly subservient to the PHOs which are the Government's preferred contract provider (Cameron 2007b). This shift has affected the morale of many IPA leaders and members, especially given their previously important role in developing financial and quality initiatives during the 1990s (Barnett 2003).

**International Comparisons**

The New Zealand experience indicates attempts by the Government to strengthen and increase the effectiveness of primary care as a key part of the health system. A comprehensive framework for effective primary health care was first set out in the Alma-Ata Declaration (WHO 1978). The Declaration incorporated two key ideas: the operational services that ensure high quality care; and the organizing principles of the system. The operational clauses reflect undisputed features of good primary care: essential services at the point of first contact, family-centred continuing care, and well coordinated care and with a preventive focus. In Western countries, these operational aspects of Alma-Ata reflected emerging trends in good practice promoted by professional associations and supported by governments. The organizing principles, however, were more contentious: population health approaches and addressing inequalities in access and outcomes; democratization of decision-making; and placing primary care in the centre of the health system. These challenged both the control over primary care exerted by GPs and the priority traditionally given to hospital and specialist services for first call on health sector resources. Despite these challenges, the lack of responsiveness to Alma-Ata principles seems surprising considering that, for example, increased emphasis on population health and prevention were expected to improve health status, help negate the adverse effects of income inequality (Hefford et al. 2005; Shi and Starfield 2000) and reduce the pressure on (expensive) hospital services (Starfield et al. 2005). Community participation was also expected to make services more responsive to local communities, thereby improving access and reducing inequalities. Making primary care the so-called centre of the health system would ensure better resourcing and a greater decision-making role for providers.

The previous discussion of primary care developments in New Zealand since 2000 reported on efforts to promote an Alma-Ata approach. It is useful to set these efforts in a wider context by comparison with other systems, for example those of Australia and the UK, on a selected range of criteria that relate closely to key themes of the Alma-Ata approach. These themes include: placing primary care at the centre of the health system; developing the organization, funding and accountability mechanisms for population health strategies and addressing inequalities; and the democratization of decision-making, both for policy and service delivery.

*Primary Care as the Centre of the Health System*

Of the three countries mentioned above, the National Health Service (NHS) in the UK was the first to declare primary care to be the centre of the health system by promoting a 'primary care led NHS' (NHS Executive 1994), using GP fundholding as the main policy instrument. This was possible because of the historical relationship between general practice and the state, characterized by the presence of patient registers and capitation payments. Primary care policy in the 1990s focused initially on achieving efficiency gains rather than equity in health status and was structured at a micro-level around practices. For example, GP fundholding was largely concerned with managing resources and the responsiveness of secondary care to patient needs, and not directed to broader population health issues (Green et al. 2006). Over the last decade, however, larger organizations have developed, first Primary Care Groups and more recently Primary Care Trusts which have a funding role and a population base. Despite the positive rhetoric surrounding Primary Care Trusts, traditional health system power relationships have tended to discourage a greater role for primary care, with secondary care providers likely to express doubts over the capability of primary care and resistance from other stakeholders to competing claims for resources (O'Cathain et al. 1999). Nevertheless, Primary Care Trusts are statutory bodies that fund both primary and secondary services and have mechanisms through which more explicit primary care policy, particularly improving access, widening choice, improving responsiveness of services and population health initiatives, can be promoted.

Despite turning its attention somewhat later to primary care, in 2001 New Zealand developed a much more explicit approach to primary care policy than other countries through the New Zealand Primary Health Care Strategy. This provided the basis for the changes outlined above, and has become the focus of attempts to improve population health status and reduce inequalities. In contrast to the NHS, a wider funding or commissioning role has not so far been considered for PHOs and so they lack the financial authority of Primary Care Trusts. Nevertheless, the linking of the New Zealand Primary Health Care Strategy to the broader public health goals of the New Zealand Health Strategy ensures that primary care is well integrated into the strategic directions for the health system.

In Australia, Government policy in relation to primary care is relatively non-specific and comprises a series of general statements such as 'making care more accessible'; 'focusing on prevention and early intervention'; and 'encouraging better management of chronic disease' (Dept of Health and Ageing 2004). As Scott and Coote (2007, 16) note, while 'there is frequent acknowledgement of the importance of a strong primary care sector, this … has never led … to the development of a defined, endorsed, national primary care strategy', although nine national priority areas have now been developed, relating to dimensions of access, clinical management and infrastructure/process. Government action has affirmed the private nature of primary care and not intervened to create structural change, precluding a systematic population approach to addressing health inequalities. Internationally, social insurance and fee-for-service models are not always well integrated with other parts of health systems, do not encourage delegation (Gress et al. 2006) and rank low on team work (Boerma 2006), all traditional criticisms of the commercial independent practice model found in Australia (Scott and Coote 2007).

*Funding, Organization and Accountability Mechanisms*

Not unexpectedly, with differing policy approaches to primary care in the three countries, there is also variation in the funding, organizational and accountability mechanisms available to address inequalities. In New Zealand, population-based provider organizations are accountable for implementing government strategies. PHOs have been the mechanisms for bringing GPs into a closer relationship with the state and for supplying the levers necessary to tackle inequalities in access and health status. The introduction of capitation funding, very low access supplementary payments and the limitations placed on co-payments have all contributed to lowering the costs of access. Accountability through a contractual relationship with the funder and the provision of incentives through the Performance Management Programme (PMP) now give the opportunity to hold PHOs and practitioners accountable for, and to reward, performance. The PMP uses indicators that are both sensitive to health inequalities and designed to incentivise the targeting of particular service strategies. However, it remains limited with few indicators and limited rewards, with most quality initiatives reliant still on professional incentives.

The recent nature of these initiatives contrasts with the more developed situation in the UK where there has been a longer tradition of contractual relationships between GPs and the state, and free consumer access via capitation funding. The presence of a commissioning role has enabled Primary Care Trusts both to expand the range of services and ensure service responsiveness (Smith and Mays 2007). Furthermore, a more comprehensive approach to performance management has occurred through the introduction of a comprehensive Quality and Outcomes Framework (QOF) which provides significant financial incentives directly to practitioners for performance. A recent analysis of the QOF (Roland 2007) concludes that its introduction coincided with a decade of progress in quality and infrastructure development, explaining the ease with which practitioners achieved targets. Initial experience suggested that QOF incentives were actually widening inequalities, a perverse outcome; but the weight of analysis now suggests that in fact there is relatively little difference in the performance of practices in affluent and deprived areas (Doran et al. 2006). However, the impact on professional values has been uncertain (Roland 2007), suggesting the need for continuing reappraisal of the system.

In Australia, the persistence of an individual relationship between GPs and the state through Medicare social insurance funding has maintained the fee-for-service arrangement, without significant structural change. This approach ensures relatively modest levels of co-payments and good access to care for much of the population but does not address population inequalities, although an increasing number of financial incentives are available through Medicare for preventive work and to address rural and indigenous health needs. Particularly interesting is the way in which some funds are now channeled through an important organizational innovation, the Divisions of General Practice. The Divisions are voluntary network organizations that now have over 90 percent of GPs as members. Since 1992 the Commonwealth government has provided significant financial support for the Divisions, initially to develop practice infrastructure, enhance practitioner education and promote quality improvement and a more integrated approach to primary care. More recently Divisions have been funded to ensure the delivery of a range of programmes through general practice, including immunization, health promotion and nutrition, and after hours and locum services. A National Quality and Performance System for Divisions of General Practice was introduced in 2005. Research evaluation based on reports from Chief Executive Officers of Divisions suggests that while the congruence between local programmes and national objectives is generally good, there is some disquiet that the focus on indicators may divert resources away from locally determined priorities (Gardner et al. 2006).

*Democratization of the System*

Democratization of primary care can occur at two levels: through the involvement of the community and consumers in decision-making and through the changing relationships between the state, doctors and other health professionals. In relation to the community as a whole, the democratization of primary care organizations has gone farthest in New Zealand, with PHOs being structured as non-profit non-government organizations required to have representatives of the wider primary care sector, the community as a whole and Maori. The third sector approach therefore has become the dominant model in New Zealand, in contrast to the UK, where PCTs are statutory entities and strongly provider driven, or the Divisions of General Practice in Australia which are provider controlled. There are moves in the UK to encourage alternative social enterprise models to provide opportunities for more diverse, community controlled and population oriented services (Lewis et al. 2006). It is likely that issues of health inequality will be better tackled by emphasizing the integration of public health and primary care (Iliffe and Lenihan 2003).

   At a consumer level, Australia has a particularly strong history of consumer involvement in health services through state-consumer health councils and, in primary care, a consumer focus is part of the latest contract between Divisions and the government, with Divisions required to ensure effective involvement. Sixty-five percent of Divisions had programmes to improve collaboration with consumers, mostly through reference and advisory groups, with 61 percent reporting formal mechanisms to involve indigenous organizations or consumers (Scott and Coote 2007, 21). Similar approaches are used in New Zealand, whereas in the UK responsiveness to the declared problems of consumers, such as waiting times and appointments has become part of significant policy initiatives and performance accountability. In general, particularly in New Zealand, the growth of consumerism has an impact on the democratization of provider systems, as community organizations seek greater control over the provision of health services.

In the case of democratization of providers, new collective organizations such as IPAs, Primary Care Groups and Divisions provided opportunities to broaden the scope of general practice and create a stronger primary care movement. These moves were also consistent with changed political priorities and the desire of governments in all three countries to strengthen primary care and widen its scope beyond that of general practice. Such moves inevitably encouraged the apparatus of the state to become more involved with primary care and perhaps undermine the autonomy of practitioners. In the UK and New Zealand it has been most noted that broadened responsibilities have affected the autonomy of GPs by increased accountabilities to public funders and the wider community.

There is now also greater involvement of other health professionals, with other health providers undertaking some of the tasks previously undertaken exclusively by GPs. In all three countries increasingly, nurse practitioners and other health workers are able to conduct their work outside doctors' control; this includes the new Family Health Nurse position in Europe (WHO 2000). Admittedly, independent nursing roles are less well developed in New Zealand and, particularly, Australia (Halcomb, Patterson and Davidson 2006).

## Conclusion

Change in primary care systems has varied substantially between countries. While a full understanding of the nature and extent of such changes is beyond the scope of this chapter, it is clear that they have affected the professional autonomy of health providers and had important implications for the costs, effectiveness and equitable provision of primary care services and primary health care more broadly. This has been particularly true in New Zealand which has seen a change in power relationships between the medical profession and the state, which has been quite remarkable given the longstanding adversarial relationship between these two groups. The fact that the dominant fee-for-service model of earlier years has been replaced by a third sector community-oriented model is not only theoretically interesting but also is extremely significant for disadvantaged populations. In one sense the state, in attempting to implement a new model of care, was able to capitalize upon the uncertainties of general practice in New Zealand and to offer GPs significant monetary inducements in order to implement change. To put it crudely it could be said that GPs 'were bought off' as part of the re-orientation of funding in the health sector. Yet to attribute such change to only top-down factors would be an oversimplification as changes in primary care were also to some extent driven by factors internal to general practice itself and the advent of IPAs, formal organizations with which the government could negotiate. This too perhaps paved the way for more successful state intervention to reorient the profession towards government equity objectives which had been largely ignored by the IPA professional collectives developed in the 1990s. For the above reasons, the change toward a more community oriented, and a less doctor driven, primary care sector appears to have proceeded furthest in New Zealand compared to the other two countries discussed above. However the long-term benefits of such a strategy in terms of reduced inequality or decreased levels of hospitalization, for example, remains to be seen.

Given the importance of primary care in modern health systems it is important that future research pay more attention to the organizational forms which have evolved in different national contexts and to the political, economic and social forces which have shaped the evolution of those forms. While there has been much interest in the contextual determinants of health, geographers have paid less attention to contextual determinants which have shaped health service change. This is an important oversight and one that needs to be rectified if we are to understand processes of health restructuring and the wider implications of such changes for tackling inequalities in health and health care.

## References

Baker, D. and Hann, M. (2001), 'General Practitioner Services in Primary Care Groups in England: Is There Inequity Between Service Availability and Population Need?', *Health & Place* 7:2, 67–74.

Barnett, J.R. (1993), 'Does the Geographic Distribution of Physicians Reflect Market Failure? An Examination of the New Zealand Experience, 1981–87', *Environment and Planning A* 25, 3–22.

Barnett, J.R. et al. (1998), 'Declining Professional Dominance?: Trends in the Proletarianisation of Primary Care in New Zealand', *Social Science & Medicine* 46:2, 193–207.

Barnett, J.R. (2001), 'Coping with the Costs of Primary Care?: Household and Locational Variations in the Survival Strategies of the Urban Poor', *Health & Place* 7, 141–57.

Barnett, J.R. and Barnett, P. (2004), 'Primary Health Care in New Zealand: Problems and Policy Approaches', *Social Policy Journal of New Zealand* 21, 49–66.

Barnett, J.R. and Brown, L.J. (2006), '"Getting into Hospitals in a Big Way": The Corporate Transformation of Hospital Care in Australia', *Environment and Planning D: Society and Space* 24, 283–310.

Barnett, J.R. and Kearns, R.A. (1996), 'Shopping Around? Consumerism and the Use of Private Accident and Emergency Clinics in Auckland, New Zealand', *Environment and Planning A* 28, 1053–75.

Barnett, J.R. and Lauer, G. (2003), 'Urban Deprivation and Public Hospital Admissions in Christchurch, 1990–1997', *Health and Social Care in the Community* 11, 299–313.

Barnett, J.R., Coyle, P. and Kearns, R.A. (2000), 'Holes in the Safety Net? Assessing the Effects of Targeted Benefits Upon the Health Care Utilisation of Poor New Zealanders', *Health and Social Care in the Community* 8, 1–13.

Blendon, R.J. et al. (1998), 'Understanding the Managed Care Backlash', *Health Affairs* 17:4, 80–94.

Boerma, W. (2006), 'Coordination and Integration in European Primary Care', in R. Saltman, A. Rico and W. Boerma (eds), *Primary Care in the Driver's Seat? Organisational Reform in European Primary Care* (Buckingham: Open University Press), 3–21.

Boerma, W. and Dubois, C. (2006), 'Mapping Primary Care Across Europe', in R. Saltman, A. Rico and W. Boerma (eds), *Primary Care in the Driver's Seat? Organisational Reform in European Primary Care* (Buckingham: Open University Press), 85–104.

Cameron, A. (2007a), 'Encounters with Bean Counters. Tales of Fees Reviews', *New Zealand Doctor* August 15, 10–11.

Cameron, A. (2007b), 'Ministry Confirms it Wants IPAs Out', *New Zealand Doctor* August 29, 2007, 3.

Christchurch Press (2007), 'Anger at National's Plan to Boost Doctor Fees', *The Christchurch Press* September 27, A1.

Crampton, P., Davis P. and Yee-Lay, R. (2004), 'Comparisons of Private For-profit with Private Community Governance Not-for-profit Primary Care Services in New Zealand', *Health Services Research and Policy* 9:S.2, 17–22.

Crampton, P., Dowell, A.C. and Woodward, A. (2001), 'Third Sector Primary Care for Vulnerable Populations', *Social Science & Medicine* 53:11, 1491–1502.

Davis, P. (1985), 'Office Encounters in General Practice in the Hamilton Health District 1: Social Class Patterns Among Employed Males', *New Zealand Medical Journal* 98, 792–8.

Davis, P. (1986), 'Office Encounters in General Practice in the Hamilton Health District 3: Social Class Patterns Among Females', *New Zealand Medical Journal* 99, 573–6.

Doran, T., Fullwood, C. and Greville, H. et al. (2006), 'Pay for Performance Programs in Family Practice in the United Kingdom', *New England Journal of Medicine* 355, 375–84.

Gardner, K., Longstaff, D. and Sibthorpe, B. (2006), *Impact of the Introduction of National Performance Indicators on General Practice* (Australian Primary Health Care Research Institute, Australian National University, Canberra).

Gress, S., Delnoij, D. and Groenewegen, P. (2006), 'Managing Primary Care Behaviour', in R. Saltman, A. Rico and W. Boerma (eds), *Primary Care in the Driver's Seat? Organisational Reform in European Primary Care* (Buckingham: Open University Press), 184–200.

Gribben, B. (1993), 'Satisfaction with Access to General Practitioner Services in South Auckland', *New Zealand Medical Journal* 106, 360–2.

Gribben, B. and Cumming, J. (2007), *Evaluation of the Primary Health Care Strategy: Practice Data Analysis 2001–2005* (Auckland: CBG Health Research Limited).

Halcomb, E.J., Patterson, E. and Davidson, P.M. (2006), 'Evolution of Practice Nursing in Australia', *Journal of Advanced Nursing* 55:3, 376–88.

Hefford, M, Crampton, P. and Foley, J. (2005), 'Reducing Health Disparities Through Primary Care Reform', *Health Policy* 72, 9–73.

Iliffe, S. and Lenihan, P. (2003), 'Integrating Primary Care and Public Health: Learning from the Community-oriented Primary Care Model', *International Journal of Health Services* 33:1, 85–98.

Johnston, G. and Wilkinson, D. (2001), 'Increasingly Inequitable Distribution of General Practitioners in Australia', *Australian and New Zealand Journal of Public Health* 25:1, 66–70.

Kearns, R.A. and Barnett, J.R. (1997), 'Consumerist Ideology and the Symbolic Landscapes of Private Medicine', *Health & Place* 3, 171–80.

Lewis, R., Hunt, P. and Carson, D. (2006), *Social Enterprise and Community-based Care: Is There a Future for Mutually-owned Organisations in Community and Primary Care* (London: King's Fund), 24.

Malcolm, L. (1996), 'Inequalities in Access to and Utilisation of Primary Care Services for Maori and Low Income New Zealanders', *New Zealand Medical Journal* 109, 356–8.

Malcolm, L. (1998a), *Primary Care Utilisation and Expenditure in the Auckland Subregions* (Auckland: Health Funding Authority).

Malcolm, L. (1998b), 'Capitated Primary Care: Policy, Problems and Prospects', *Health Manager* 5, 7–9.

Matheson, D.P. and Hoskins, R.S. (1992), 'The General Practice Scheme: Was it Targeted?' *New Zealand Medical Journal* 105, 35–36.

Macintyre, S., Ellaway, A. and Cummins, S. (2002), 'Place Effects on Health: How Can we Conceptualise, Operationalise and Measure Them?', *Social Science & Medicine* 55, 125–39.

Meads, G. et al. (2006), 'The Management of New Primary Care Organisations: An International Perspective', *Health Services Management Research* 19, 166–73.

NHS Executive (1994), *Developing NHS Purchasing and Fundholding: A Primary Care-led NHS* (Leeds: Department of Public Health).

O'Caithin, A., Musson, G. and Munro, J. (1999), 'Shifting Services from Secondary to Primary Care: Stakeholders Views of the Barriers', *Journal of Health Services Research and Policy* 4, 154–60.

Roland, M. (2007), 'The Quality and Outcomes Framework: Too Soon for a Final Verdict', *British Journal of General Practice* 57, 525–7.

Scott, A. and Coote, E. (2007), *The Value of the Divisions Network: An Evaluation of the Effect of Divisions of General Practice on Primary Care Performance* (University of Melbourne, Melbourne Institute of Applied Economic and Social Research).

Shi, L. and Starfield, B. (2000), 'Primary Care, Income Inequality and Self-rated Health in the United States: A Mixed Level Analysis', *International Journal of Health Services* 30, 541–555.

Smith, J. and Mays, N. (2007), 'Primary Care Organisations in New Zealand and England: Tipping the Balance in Favour of Primary Care?', *International Journal of Health Planning and Management* 22, 3–19.

Starfield, B., Shi, L. and Macinko, J. (2005), 'Contribution of Primary Care to Health Systems and Health', *Milbank Quarterly* 83, 457–502.

World Health Organization (1978), *Primary Health Care*. Report of the International Conference on Primary Health Care, Alma-Ata, USSR, September 6–12, 1978. (Geneva: World Health Organization).

World Health Organization (2000), *The Family Health Nurse: Context, Conceptual Framework and Curriculum* (Geneva: World Health Organization).


**Internet-based References**

Barnett, P. (2003), 'Into the Unknown: The Anticipation and Experience of Membership of Independent Practitioner Associations'. *New Zealand Medical Journal* 116:1, 1171. Available at http://wwww.nzma.org.nz/journal/116-1171/382/.

Consumers' Institute of New Zealand (2007), 'GP Fees. The Big Losers in the Government's Efforts to Drive Down the Cost of Living Seem to be Young Kids'. *Consumer Magazine* May 8. Available at www.consumer.org.nz.

District Health Boards of New Zealand (2007), *PHO Performance Management Programme* (Wellington: District Health Boards of New Zealand). Available at http://www.dhbnz.org.nz.

Schoen, C. et al. (2004), 'Primary Care and Health System Performance: Adults' Experiences in Five Countries'. *Health Affairs* Web Exclusive W4, 487–503. Available at http://content.healthaffairs.org/cgi/reprint/hlthaff.w4.487v1.pdf.

Chapter 10

# New Health Geographies of Complementary, Alternative and Traditional Medicines in Primary Health Care

Daniel Hollenberg and Ivy Bourgeault

Over the last 30 years, the concept of primary health care (PHC) has increasingly become a focus across many health-related areas, including health promotion, health services, health policy and international/global health. Bridging scales, PHC has become important at patient, practitioner, institutional, governmental and, indeed, worldwide levels. As the first point of contact with any type of health care, treatment or service, PHC is also being recognized as being extremely important to all peoples, regardless of culture or background. The increased emphasis on PHC could perhaps be found in several of its ideals, such as the attempt to make health care more accessible, acceptable, equitable and affordable. Moreover, in addition to general attention to the needs of individuals and their respective communities, community participation and health education are key PHC principles.

Within the changing landscapes of PHC, the purpose of this chapter is to focus on a type of health care that is not usually thought of by the *biomedical community* as a part of mainstream PHC delivery: complementary/alternative and traditional medicines (CAM/TM). We define CAM/TM here as non-biomedical treatment modalities that are often used as direct forms of health care in different contexts, regardless of their recognition by mainstream PHC campaigns. CAM/TM has become increasingly important, if not essential, to people around the world.

In line with this volume's main theme of place and PHC, our main argument is that health care modalities drawn from CAM/TM form unique configurations of PHC in many different contexts. Earlier mainstream forms of PHC usually did not include specific links to CAM/TM; or, put differently, CAM/TM is not usually thought of as a form of PHC. With notable recent exceptions by a handful of scholars (e.g., see Bodeker 2006; Bodeker and Burford 2007), it seems that non-biomedical health modalities have gone under the 'radar' of major PHC initiatives, or have simply been regarded as avenues for future investigation without a pragmatic focus.

In this chapter, we view CAM/TM and its new interactions with and integration into PHC as global health phenomena, and simultaneously as particular and local phenomena or health geographies. We argue further that traditional and indigenous health knowledges directly link up with different types of PHC in specific geographical locations. As Andrews (forthcoming, 9) notes of the contributions of an early health geographer of TM, Charles Good, 'the research agenda Good posed was focused on accessibility and utilization of services including investigating spatial arrangements of traditional medicine in urban and rural areas, factors in seeking traditional medicine and various obstacles related to distance.' Today, health geographers have expanded Good's early focus to include 'a deeper and often qualitative understanding of the cultures, values and belief systems that relate to traditional medicine in particular places of the developing world' (Andrews forthcoming, 9). Still, even with this focus, the role of CAM/TM as a specific form of PHC in the context of place and landscape has not extensively been explored.

Below, we present a case study on an integrative health care setting combining CAM/TM and biomedical therapies, and argue for its theoretical location as one aspect of PHC. We specifically respond to Andrews' (2007) request for researching: aspects of integration within conventional medicine in shared places; how different types of CAM/TM may be integrated with conventional places for health care; and design features of CAM/TM settings, and how they play an important part in the overall CAM/TM consumer experience. We first briefly review important concepts of PHC as it applies to CAM/TM. Following this we discuss how CAM/TM and PHC have specific health geographies.

## PHC and CAM/TM: A Brief Review

In the Alma-Ata Declaration by the World Health Organization (WHO) in 1978, important principles of PHC were identified, including equity, community involvement/participation, intersectorality, appropriateness of technology and affordable costs. Drawing on these concepts, Nutbeam (1998) states that 'At a very practical level, there is great hope for both planned and opportunistic health promotion through the day-to-day contact between PHC personnel and individuals in their community' (1998, 352–353). Reflective in part of this broader health promotion and prevention focus, PHC has become much more interdisciplinary and collaborative, involving a range of practitioners and treatment modalities (Besrour 2003).

In the Alma-Ata Declaration, several important concepts are highlighted that connect to CAM/TM. For example, the Declaration states that, 'The people have the right and duty to participate individually and collectively in the planning and implementation of their health care' and that 'It is the first level of contact of individuals, family and community with the national health system, bringing health care as close as possible to where people live and work, and constitutes the first element of a continuing health care process.' Furthermore, this foundational declaration indicates that PHC involves: the appropriate treatment of common diseases and injuries; maximum community and individual self-reliance and participation in the planning, organization, operation and control of PHC; the fullest use of local, national and other available resources; use of local health workers including physicians, nurses, midwives, community traditional practitioners as needed; and responding to the expressed health needs of the community (WHO 1978).

In spite of the detail provided by the Alma-Ata Declaration, the WHO realized some years later that no universally applicable definition of PHC exists. For example, in the Alma-Ata document, PHC was discussed as both a level of care and an overall approach to health policy and service provision. Moreover, in high- and middle-income countries, PHC is understood, for the most part, as the first level of care, while, in low-income countries, PHC is largely seen as a system-wide strategy. In essence, the global, national and local environments in which PHC values have been translated into action have changed fundamentally over the last 30 years. Policy documents endorsed by the WHO (e.g., Atun 2004) have indeed noted that defining PHC is fraught with difficulties, such as the existence in the United States (US) of no fewer than 92 definitions of PHC, and the realization that within the European Union the definition of PHC varies country-by-country. As a result, PHC today can be understood along many different strata, which involve PHC as a concept, level, content of services, process and team membership (Atun 2004, 6).

In terms of PHC as a level of care, it can be understood as 'the domain where people first contact the health care system and where 90 percent of their health problems are dealt with' (Atun 1998, 16). As a related concept, PHC as a 'strategy' involves the 'notion of accessible care relevant to the needs of the population, functionally integrated based on community participation, cost-effective and characterized by collaboration between sectors of society' (Atun 1998, 16). PHC as a 'philosophy' involves the equitable delivery of health care with particular reference to 'intersectoral collaboration.' Differences between selective and comprehensive PHC have also further complicated the debate. The former refers to an essential set of health interventions, or a package of health services. The latter comprises a wide range of health services, including health education, promotion, prevention, curative and rehabilitative care.

Health care restructuring in industrialized, wealthy nations is changing the boundaries of what PHC is considered to be where, for example, the de-listing of in-patient hospital services is now shifting care to the community level. The occurrence of hidden primary care has also occurred, defined by the new adoption of PHC skill sets by certain health care practitioners who were previously not responsible for these duties. The shifting professional boundaries and role sets in PHC is further demonstrated by the wide variety of PHC teams being established, which can range from community nurse-rural general practitioner pairings up to multidisciplinary teams of 30 professions including a variety of PHC providers (Bourgeault and Mulvale 2006). Upon closer examination, tensions between regulated and unregulated PHC practitioners are also apparent. For example, a definition of a PHC professional, provided by the Royal College of General Practitioners in the United Kingdom, states that:

> A primary health care professional is any health professional whose professional qualification is in health care, whose professional qualification is recognized by a statutory registration council approved by parliament, who sees clients/patients directly without any referral from a health professional, or who worked within a primary medical or nursing care organization that offers patients open access (Atun 1998, 17).

Yet, members of PHC multidisciplinary teams are often health professions such as midwives, osteopaths and other non-regulated CAM/TM practitioners. Although they provide PHC, they are often not included in official definitions of PHC professions.

There are several important observations that can be made from the above discussion. It is clear that CAM/TM clearly fit into several dimensions of PHC, such as: using appropriate health care technology; drawing upon well-placed and local health care personnel; recognizing the right of patients/clients to choose their own care; choosing the appropriate treatment for illness and disease; and drawing on suitably trained health care practitioners that can respond to the health needs of the community. These are all important dimensions where CAM/TM is highly applicable. Furthermore, the varying definitions of PHC, the breadth of health care services to which PHC may apply and the shifting professional boundaries between both regulated and unregulated health practitioners can only further support the use of CAM/TM as viable PHC modalities. Most importantly for our discussion is the realization that PHC can occur in extremely specific and contextual locales, and that if PHC can take many forms depending on the needs of the community, then there is no question that CAM/TM is providing specific PHC services in local health geographies.

## Locating CAM/TM and PHC

In many non-Western countries where TM in particular is used, the biomedical physician-to-population ratio for rural areas is sometimes as low as one biomedical health care provider for every 100,000 citizens (WHO 2002). In these countries PHC is most often delivered as it has been for generations: in the form of traditional medicines passed down orally, often with a long history, and including both specific health care practices and indigenous knowledges. While TM is sometimes highly criticized by the biomedical community, it comprises richly detailed, locally embedded and internally consistent healing modalities that can produce pragmatic health effects for the population it is serving. It was 30 years ago that the WHO declared:

> Traditional medicine practitioners and birth attendants are found in most countries. They are often members of the local community, culture and traditions, and continue to have high social standing in many places, exerting considerable influence on local health practices. With the support of the formal health system, these indigenous

practitioners can become important allies in organizing efforts to improve the health of the community. ... It is therefore well worthwhile exploring the possibilities of engaging them in primary health care (WHO 1978).

Indeed, perhaps ironically, today we have almost come full circle. PHC initiatives used to be introduced in opposition to traditional medical systems, whereas now the new collaborative, multidisciplinary focus of PHC better enable the integration of various therapeutic systems/modalities.

Given the extremely low ratio of biomedical health practitioners to citizens in many rural non-Western nations, it seems perhaps obvious that local traditional medicines would be investigated as one of the first options or strategies in the context of international PHC campaigns. Yet, it is apparent that even despite detailed policy documents issued by the WHO regarding the usefulness and implementation of TM and related traditional health strategies (e.g., see WHO TM Strategy 2002–2005), international and mainstream PHC campaigns have, for the most part, not drawn upon or effectively utilized indigenous local knowledges. This observation is most apparent upon examination of the HIV/AIDS and related health campaigns against infectious disease, which primarily focus on access to and distribution of anti-retroviral medications while indigenous health knowledge at the local and community level has been largely ignored. Policy analyses by international leaders in TM regarding development policy continually reinforce this point (Bodeker and Burford 2007).

This is a curious reality, as, for example, one of the first things a South African patient with HIV/AIDS will initiate is a visit with a local/indigenous health practitioner.[1] Furthermore, consistent work in the area of TM and infectious disease by Bodeker and Burford (2007) and others extensively documents the global use of TM and its effectiveness, even within the context of treating specific symptoms caused by HIV and malaria. Similarly, rising trends in CAM/TM use in mostly northern 'rich' and Western industrialized countries (and across gender, age and illness-related categories) are marking CAM/TM as a definite first option form of PHC. A survey in the US, for example, revealed that a significant proportion of patients were consulting a CAM/TM practitioner as their primary care provider more often than their family physician (Faass 2001). More than ever before, CAM/TM is being used as a direct form of PHC, whether individually or in combination with conventional integrative biomedicine, and in different health locales.

There are arguably many valid and legitimate push/pull factors for this development, and they should not seem surprising, given that: the incidence and prevalence of chronic-type illnesses in rich nations (e.g., chronic pain) has

---

1  We are definitely not arguing that access to anti-retroviral medication is not essential for the survival of millions of people with HIV/AIDS, nor that these campaigns should be slowed down in any way, or that the many problems that even these mainstream PHC campaigns have recently experienced should not be resolved (e.g., the barriers to producing and manufacturing generic HIV/AIDS drugs).

far outweighed acute-type illnesses for the last 50 years, for which biomedicine has little to offer; the incidence of iatrogenic disease (i.e., disease caused by biomedicine) has risen dramatically, especially with the increased use of high-tech biomedicine; and perhaps of most importance for the patient, different types of CAM/TM practices – whether pure placebo, medicinal active ingredient, particular manipulative technique, validation and respect of a patient, or a combination of them all – have greatly ameliorated patients' suffering from these chronic-type illnesses.

One of the main differences in the recognition of CAM is that, differing from TM, the rising demand and public use of CAM has garnered the attention of both government and conventional medical schools. In the US for example, the popularity of CAM has prompted different medical and educational institutions to band together to form an academic consortium of integrative medicine (i.e., care that integrates the practices of biomedicine and CAM within the institutional confines of academic biomedicine). Their main focus is to glean the benefits of various CAM modalities, to simultaneously train physicians in these new holistic therapies and to integrate both into new medical school curricula and residencies. Called *integrative medicine* (IM) or *integrative health care* (IHC), hundreds of new types of health care settings have emerged in biomedically-dominated systems, where various forms of CAM are combined with aspects of conventional biomedicine in both clinics and hospitals. Interestingly, the recent expansion of medical school enrolment in the US has been particularly salient in the more integrative osteopathic colleges (Croasdale 2006).

## IHC Settings as New Geographies of PHC

IHC settings have become increasingly present over the last 15 years, thereby providing a more structured health care service for those individuals seeking CAM/TM modalities. These settings provide a new space in which patients can explore multiple healing modalities under one roof. Due to the newness of these settings and also their sometimes parallel features with other biomedical health care settings such as community health care centres, variations in the concept, philosophy and structure of IHC settings vary significantly in the health policy literature.

Although many IHC concepts remain debatable, certain parameters can be drawn here. It is generally agreed upon that IHC settings involve a varied mix of either purely CAM/TM, or CAM/TM and biomedically-related health care modalities. Within the CAM/TM professions alone (e.g., chiropractic, massage therapy, traditional Chinese medicine and naturopathic medicine, among many others), literally hundreds of what can be termed *holistic* or CAM/TM-type IHC clinics exist throughout many Western countries, that offer multidisciplinary care to the community. Within CAM/TM settings, the main concept of integration refers to how the various CAM/TM professions interact, both clinically and professionally, to combine and integrate their respective services.

The second main type of IHC setting, which can be called CAM/TM/biomedical IHC settings, are defined by a relatively new trend where various CAM/TM and conventional biomedical therapies and professions are combining their services in a single clinical setting. For example, this second type of CAM/TM/biomedical IHC setting also involves a mix of various CAM/TM professions (as noted above), but may also involve such biomedical and paramedical practitioners as family physicians, nurses and physiotherapists. At present, the number of holistic CAM/TM-IHCs is significantly higher than the number of IHC settings combining both CAM/TM and biomedical health care practitioners.

As noted, there is a recent trend in the biomedical community to begin to incorporate various CAM/TM modalities into a form of integrative medicine. It is now commonly noted, however, by critically informed health policy theorists and analysts that the move to incorporate CAM/TM into biomedicine most obviously reflects: yet another professional takeover by biomedicine of competing health care professions (such as was done historically in the establishment of medical dominance); the biomedical attempt to *increase* patient flow due to the popularity of CAM/TM services; the recognition by large-scale biomedical institutions that CAM/TM is generating significant billing dollar revenue; and in general, the dramatic rise in popularity of the use of CAM/TM as a form of PHC (Baer 2004; Hollenberg 2006a, 2007; Saks 1994). Despite biomedical attempts at CAM/TM appropriations, the relatively low number of CAM/TM/biomedical IHC settings in comparison with CAM/TM-IHC settings reflect the observation that CAM/TM and biomedical therapies still operate from sometimes radically different therapeutic paradigms. Moreover, different sectors of the biomedical community remain highly critical of the use of CAM/TM to treat illness and disease.

It should be noted that various types and forms of CAM/TM/biomedical IHC occur that do not always involve a single physical or geographical space and/or location. For example, patients are already combining various forms and types of CAM/TM and biomedicine as a form of integrative care, and many solo physicians are incorporating CAM/TM into their clinical practice, but well outside of a richly detailed and complex physical space.

Within the geographical health care settings that occur in CAM/TM/biomedical IHC, by far the most complex are interdisciplinary integrative group practices. Other less detailed spaces, however, also exist in the form of multidisciplinary integrative group practices who may share the same clinic space; hospital-based integrative care; and integrative medicine in an academic medical centre.

In a multidisciplinary integrative group practice, conventional and CAM/TM practitioners work together in the same centre or clinic, share physical space, cross-refer and may also focus on a particular illness or disease (Mann et al. 2004). Biomedical and CAM/TM practitioners may see the same patients, but following an interdisciplinary model they would do so separately and without necessarily convening in a formal or group team meeting. Obviously, the main advantage of a multidisplinary IHC model is the ability to offer comprehensive biomedical and CAM/TM services to a particular patient population. As all practitioners would be working in the same physical location, there is great potential for information sharing amongst different therapeutic disciplines, which would ultimately benefit practitioners and patients. Within hospital-based integrative care, aspects of various CAM/TM modalities and conventional medicine are delivered within the same hospital system or medical centre. Although delivered within the same physical location, to date most CAM/TM services within hospital-based integrative care are delivered through an out-patient hospital service, which is usually far removed both clinically and geographically speaking from conventional hospital care.

In contrast to multidisciplinary and hospital-based integrative care, interdisciplinary integrative group practices attempt to achieve a high level of integration. In addition to CAM/TM and biomedical practitioners sharing the same physical geographical location and office space, ideally they share common patient records and also regularly hold group meetings to discuss individual cases to further develop integrative care plans. As with multidisciplinary integrative group practices, interdisciplinary group practices may also focus on a specific illness or disease. Advantages for patients also include the provision of a broad spectrum of CAM/TM and biomedical services. The philosophical and geographical designs of interdisciplinary, integrative group practice spaces also provide all practitioners with in-depth learning opportunities, for example, within the integrative group rounds or meetings. One of the key geographical features of interdisciplinary integrative group practices is the provision of physical settings, or locales, that follow a holistic orientation to health care and healing, in contrast to what would be found in a conventional biomedical setting.

In sum, despite differences in types of IHC settings, these settings as a whole are now providing new types of health care locales for PHC. Shifting models of PHC are now leading to new definitions of PHC, such as care that is required, needed and chosen by various communities and not only the very first experience with a health care system. The recent proliferation of IHC settings may prove to be extremely important and relevant to policy-makers implementing new PHC models at institutional and governmental levels.

### Case Study: A Description of an IHC Setting as a PHC Locale

The description that follows largely draws upon the first author's in-depth ethnographic research within one Canadian CAM/TM/biomedical IHC setting focusing exclusively on cancer care. To further locate this setting, the below IHC site is an interdisciplinary integrative group practice (Mann et al. 2004).

*Background*

The Centre was founded in 1996 as a non-profit society funded largely by private donations, and partially by a provincial Medical Services Plan (MSP). The Centre's main mission is to provide an integrated complementary cancer care programme for people with cancer and their families, and to encourage a holistic healing approach to cancer care that includes self-care, nutrition, vitamins, supplements, and a range of complementary medical therapies. The Centre is one of the few integrative cancer clinics in Canada, and among the first to have physicians funded by a provincial government to provide complementary care. It is thus a unique IHC setting, established ostensibly on CAM principles to deal with a chronic disease that is a challenge to biomedicine.

*Practitioners*

The Centre consists of five medical doctors (MDs) and eight CAM/TM practitioners. The MDs can be classified technically as general practitioners or family physicians. They are unusual, however, in that most have a philosophical orientation and pragmatic or personal experience with various non-biomedical paradigms of healing. These unconventional interests range from nutrition, supplements, clinical hypnosis and mind/body medicine (or psycho-neuro-immunology), to acupuncture, T'ai chi, Qi gong, yoga, nature therapy and spiritual healing. For example, two of the MDs, one of whom is the Centre's director, have visited and studied in South America, Africa and Asia. The Centre's eight CAM/TM practitioners, called associate practitioners (APs), include a massage therapist, doctor of traditional Chinese medicine, registered nutritional consultant, naturopathic doctor, two homeopathic practitioners, holistic counsellor, and an integrative body work counsellor.

*Physical Layout*

The Centre is located on the second floor of a newly-built building that houses other offices. Upon entering the front door, one is immediately presented with black and white photographs of the Centre's staff and practitioners, located at the front of the reception desk – in contrast to the typical biomedical practice in Canada. The front desk is staffed by one receptionist and two other staff who perform various administrative activities, such as booking patients. Immediately to the right of the front desk is what is called 'the living room' which serves as the patient waiting room. The most welcoming room of the Centre, the living room has four couches and two chairs, several plants, one video terminal, one computer, several lamps, and a large library of donated books, tapes and videos. As opposed to a traditional medical waiting room, the living room welcomes patients into a relaxing atmosphere in which to wait for appointments.

Immediately adjacent to the living room is the kitchen. The kitchen, donated by various sponsors, provides a practical classroom for the cooking classes offered by the Centre. Across from the kitchen, along a long hallway is the 'intravenous room.' The intravenous room, filled with comfortable chairs and lit by natural light from two glass walls, provides a relaxing environment where patients receive hydrogen peroxide, vitamins C and B, ozone, and other intravenous injections and/or infusions administered by the naturopathic doctor. Three offices along the corridor house the four medical doctors, while the rest house the APs. The fifth MD works outside the Centre in a separate but related clinic. The geographical dominance of biomedical over CAM offices has been noted in other research (e.g., see Shuval et al. 2002). It could be argued that a similar geographical dominance was present here as well, in that the first offices encountered were those of biomedical practitioners. Yet overall, a 'holistic' feel to the space of the Centre was readily apparent.

*Funding Structure*

The Centre, at the time of research, had 2.5 of the five MDs funded by the provincial MSP. The MSP is the province's funding organizational structure that provides for all medical services covered under the Canada Health Act. The MDs receive a salary from the MSP for the specific purpose of extended patient visits of approximately one and a half hours. Accordingly, patients are not charged for visits with MDs. This places biomedicine in a clearly favoured position as CAM in Canada has always existed in the 'private' sector of funding (i.e., out-of-pocket or private health insurance). The other half of funding for the MDs is provided by private donations. All the MDs work part-time at the Centre – approximately two to four days a week – to accommodate the funding structure. Private donations also account for the salaries of administrative and research staff, the Centre's library and kitchen, and help to subsidize the two-day introductory programme for patients and their families.

APs rent office space in the Centre, where they locate their private practices. APs charge on a fee-for-service basis set by the individual practitioner. Several of the APs and three of the MDs also work simultaneously outside of the Centre in their own additional private practice. For example, one of the MDs worked in her own private pain clinic, while the doctor of traditional Chinese medicine was located outside of the Centre in her own private practice. The fact that CAM was strictly located in the private sector had clear consequences for patients and CAM practitioners at the Centre, including shorter treatments for patients and a lower salary for CAM practitioners (see Hollenberg 2007 for a more detailed discussion of the effects of private-sector CAM on patients and providers).

*Programmes and Activities*

The Centre offers a two-day introductory programme for patients and their families, private patient consultations with MDs (free of charge), consultations with APs and seven ongoing educational classes. The introductory programme consists of 12 hours of seminars and workshops over a two-day period, including an introduction to: complementary cancer care and healing; meditation; nutrition; visualization; group sharing; decision making; vitamins and supplements, in addition to the opportunity to meet informally with the Centre's MDs and APs. The programme also includes and invites family members or other persons to support the participant and acquire information during the seminars. For example, one seminar programme consisted of 20 participants and 13 family or friend support members. As of 2002, the programme occurred every two weeks with a subsidized cost of $280 Canadian, and $50 extra per support person, which included a two-month supply of vitamins with the retail value of $200 Canadian. Few family practices, or even holistic centres, are able to match this level of subsidization for new patients.

As noted, the Centre's MDs usually conduct private consultations of approximately one and a half hours with patients, after patients have completed the introductory programme, in order to 'fine tune' their personal cancer care programmes. Again, this is unusual in primary care in Canada, with patient consultations usually lasting no longer than ten minutes. Patients have the option to consult with MDs with or without programme completion.

Patients' private consultations with MDs involve the physicians taking a thorough medical history and reviewing current medications. This is done in order to exclude contraindications between vitamins or supplements and pharmaceutical drugs. Practitioners explained that in addition to ensuring a vitamin/supplement regime most appropriate for the patient, an appropriate follow-up plan is also made, including more in-depth consultations if necessary. Patients are encouraged to bring family members or other support persons with them to the physician consultation, which is also not typical in biomedical primary care. Patients at the Centre may choose to work with and pay privately for one or more of the eight APs by contacting them directly. The Centre's eight ongoing classes include meditation, yoga, nutrition and cooking, Qi gong, support groups, and two classes of homeopathy.

*The Centre's Philosophy*

Although biomedical dominance was subtly apparent, the Centre's philosophy is that 'illness is not separate from self' and that 'mind is not separate from body.' This differs from conventional and historic biomedical theory, where illness is usually separated from self/mind/spirit. The Centre is described as viewing how one feels as a crucially important element in the healing process. While the Centre's literature certainly recognizes the merits of conventional biomedical cancer treatment, it presents biomedicine as just one 'spoke in the wheel' of an integrated approach to healing. Specifically, in the philosophy outlined in the patient handbook, called the *Blue Book*, conventional biomedical therapies are represented as only one 'spoke' in a group of 13 others that include: supporting the will to live; spiritual and emotional connection with others; healthful diet and water; exercise; and adequate sleep and relaxation.

The texts distributed by the Centre emphasize that all spokes are interrelated and contribute to the overall health of the patient, in that the total treatment is far greater than the sum of individual therapies. This is reminiscent of the systems theory thinking that pervades the IHC literature; however, here it may be more applicable to patients rather than professions. Further, the Centre's literature presents healing as a journey that is unique to each individual. It is recognized that what is of great benefit to one person may not necessarily be of benefit to another. In this sense, the Centre suggests that patients, in addition to receiving some professional guidance, reclaim their own inner wisdom and healing, and feeling of control, in order to design their own healing recovery plans.

*Philosophical Treatment Model*

Drawing on their holistic philosophy, the Centre has expanded the conventional biomedical tumour-centred model of cancer to create what they term a new person-based model that includes the 'spokes in a wheel' concept described above. The Centre rejects the 'old model' based on the conventional options of surgery, radiation and chemotherapy in the treatment of cancer. The 'old model' is viewed as being delivered through a patriarchal and hierarchical doctor-patient relationship where patients are told what treatments they are to have, resulting in loss of patient autonomy, empowerment and self in the healing recovery process. The Centre's *new* person-based model thus acknowledges surgery, radiation and chemotherapy alongside the other 'spokes' such as diet, prayer and exercise, where patient empowerment is honoured and supported in a three-way patient-doctor relationship. The self plays a fundamental role in healing and recovery, resulting in the enhancement of the patient's well-being and immune system and the support of the mind, body and spirit in the healing process.

Despite philosophies that postulate the equality of the spokes in the wheel and the rejection of the tumour-based model, the Centre views particular spokes as more fundamental than others in terms of creating what the Centre calls a 'specific foundation of recovery', beginning with the will to live, hope and spiritual connection. Thus there is a commitment in the Centre's discourses to what they term holism, which runs counter to reductionist biomedical discourses.

## Discussion: IHC Settings

The holistic philosophy of the Centre, combined with the specific geographical layout, provided patients and practitioners the opportunity to experience particular aspects of integrative PHC. New types of geographical spaces directly enhanced this integrative philosophy and enabled a broader range of programs and activities. For example, the 'living room-type' atmosphere of the waiting room provided a physical and social environment in which patients, practitioners, family members and other parts of the community could formally and informally interact and socialize. Williams (1998) also found that places such as waiting rooms have the potential to enhance the healing process through a strong 'sense of place.' One of the patient treatment rooms with access to natural light, and also furnished with comfortable chairs, was specifically designed to relax patients during intravenous treatments. The additional existence of a kitchen and yoga/meditation area also provided physical spaces within which to enact the Centre's integrative philosophy of healing. These spatial aspects of care at the Centre are consistent with Andrews' (2003) earlier work on older people's use of CAM, where group CAM practices were described as 'hubs' of therapeutic activity. Both practitioners and patients could often be found in the meditation room, for example, participating in and/ or leading various therapeutic classes. As another way of supporting a type of integrative community bond, practitioners at the Centre often started their day within the meditation room by joining in and individually leading a group meditation practice before their professional work day began.

The philosophical, clinical and geographical aspects of an interdisciplinary integrative group practice clearly contrast with the conceptual and geographical space of, for example, hospital-based integrative care. In comparison with an out-patient acupuncture service located within a hospital in which the first author also conducted in-depth ethnographic research, there was a clear lack of integrative geographical philosophy (Hollenberg 2006b). For example, the patient waiting area within this out-patient CAM clinic was arranged in a traditional hospital-style with molded fabric chairs and fluorescent lighting, popular magazines and sparse biomedical décor. The CAM out-patient service was located in a self-contained suite with four treatment rooms and one small office, clearly separated from the other biomedical departments. It was clearly not in the consciousness of this out-patient CAM clinic, and also not within the parameters of the hospital administration, to provide other spaces related to an integrative care philosophy, such as a meditation room. Furthermore, the separation of biomedical and CAM practitioners from an integrative group philosophy could also be seen to have shaped the geographical layout of the out-patient hospital CAM clinic and the surrounding biomedical offices, which were – in typical hospital style – self-contained units separated and 'hidden' behind sterile floors and walls.

As noted earlier, this kind of geographical marginalization of out-patient hospital CAM services has also been extensively documented by ethnographic/ sociological studies of out-patient CAM clinics in publicly-funded hospitals in

Israel (e.g., see Shuval et al. 2002). In contrast to what one would expect to find within an integrative philosophy, out-patient hospital CAM clinics (in comparison with interdisciplinary integrative group practices) are almost always located in a position of geographical marginalization or physically within the 'dungeons' of the hospital.

## Conclusion: Revisiting PHC

Drawing from the above discussions, we suggest that the role of CAM/TM as an essential aspect of PHC can no longer afford to be ignored. Moreover, as the case study demonstrates, CAM/TM is used in specific places or health locales and as such, form important new health care geographies for future study.

As briefly summarized above, there are important aspects of CAM/TM that directly match up with PHC ideals in separate but clearly related ways. The moral and ethical right to plan and choose one's own health care, regardless of economic or political status, is reflected in the patient movement to purposely choose CAM/TM services over and above biomedical options. This right to choice in health care is clearly reflected in chronically ill patient populations such as the above-noted cancer population, up to 40 percent of whom choose CAM/TM options as a part of their personal commitment to ideals of holistic health (Ernst and Cassileth 1998).

Responding to community needs is reflected in the diverse number of CAM/TM and CAM/TM/biomedical clinics and centres that recognize the significant patient movement to use CAM/TM, and in turn, have created spaces for patients to choose IHC services. As noted in the case study, the cancer patient community, on the whole, clearly want access to CAM/TM and find a sense of deep satisfaction when able to attend a Centre that supports their choice in health care (Hollenberg 2006b). As a related aspect, the operation of the Centre in a residential community follows the PHC principle of providing geographically accessible health care services to the community, as do many other IHC clinics and centres.

The provision of appropriate treatment in PHC is one of the major strengths of CAM/TM, as CAM/TM is often directly tailored to the needs of the individual. In CAM/TM, diverse health care modalities are available that can respond to issues in health and wellness along physical, mental, emotional, spiritual and other health dimensions. In addition to patient preferences and appropriateness of CAM/TM for the individual, CAM/TM is being shown *clinically* to be more appropriate for specific illnesses and diseases than certain biomedical treatments. For example, specific aspects of traditional Chinese medicine (e.g., herbs and acupuncture) are being recognized as essential for cancer symptom management and rehabilitation after biomedical treatment, in place of biomedical drugs (Hui et al. 2006; Hollenberg, Cowan and Wright 2007).

Finally, the use of CAM/TM directly fits within the PHC concept of using local health care resources and practitioners. As noted earlier, CAM/TM forms a part of a cultural health care landscape characterized by readily available local health resources, such as herbal medicines and community practitioners. Although biomedicine also stems from a cultural landscape of sorts, its location begins from an *external* versus *internal* perspective in that biomedical treatment consistently operates from a non-local standpoint (e.g., synthesizing drugs from indigenous herbs [non-local standpoint] versus prioritizing community healers and herbal medicines over biomedical doctors and pharmaceutical drugs [local standpoint]) (Harding 1998). The use of local CAM/TM resources such as home herbal gardens (see Bodeker and Burford 2007) and supplements can further be seen as an aspect of the 'folk' and 'popular' sector in health care (Kleinman 1988). For example, in the case study presented above, the use of popular-sector health resources is a significant aspect of the experience of cancer patients, who often make educated decisions for the incorporation of specific dietary supplements into their treatment plan. The use of local CAM/TM practitioners at the Centre, drawn from the surrounding community, was also prevalent and further reflects a key principle of PHC.

One reason as to why CAM/TM is consistently overlooked in the development of PHC clearly relates to notions of biomedical dominance and what is conservatively thought of as 'appropriate' care. These and other tensions will have to be addressed for the wider recognition of CAM/TM as legitimate PHC practice.

### References

Andrews, G. (2003), 'Placing the Consumption of Private Complementary Medicine: Everyday Geographies of Older People's Use', *Health & Place* 9, 337–49.

Andrews, G. (forthcoming), 'Complementary and Alternative Medicine', in *International Encyclopedia of Human Geography* (Sage Publications).

Baer, H.A. (2004), *Toward an Integrative Medicine: Merging Alternative Therapies with Biomedicine* (Walnut Creek: AltaMira Press).

Besrour, S. (2003), *Colloquium: The Team in Primary Care: A New Vision, New Ways to Work*. Record of proceedings, Montreal February, Bibliothèque Nationale du Québec.

Bodeker, G. (2006), 'Global Dimensions', in M.S. Micozzi (ed.), *Fundamentals of Complementary and Integrative Medicine*, 3rd edition (New York: Saunders Elsevier), 82–92.

Bodeker, G. and Burford, G. (eds) (2007), *Traditional Complementary and Alternative Medicine: Policy and Public Health Perspectives* (Oxford: Imperial College Press).

Bourgeault, I. and Mulvale, G. (2006), 'Collaborative Health Care Teams in Canada and the U.S.: Confronting the Structural Embeddedness of Medical Dominance', *Health Sociology Review* 15:5, 481–95.

Ernst, E. and Cassileth, B.R. (1998), 'The Prevalence of Complementary/ Alternative Medicine in Cancer: A Systematic Review', *Cancer* 83, 777–82.

Faass, N. (ed.) (2001), *Integrating Complementary Medicine into Health Systems* (Gaithersburg: Aspen Publishers Inc.).

Good, C. (1977), 'Traditional Medicine: "An Agenda for Medical Geography"', *Social Science & Medicine* 11, 705–13.

Harding, S. (1998), *Is Science Multicultural?: Postcolonialisms, Feminisms, and Epistemologies* (Bloomington and Indianapolis: Indiana University Press).

Hollenberg, D. (2006a), 'Uncharted Ground: Patterns of Professional Interaction Among Complementary/Alternative and Biomedical Practitioners in Integrative Health Care Settings', *Social Science & Medicine* 62:3, 731–44.

Hollenberg, D. (2006b), *Integrative Health Care: A Critical Analysis of the Integration of Complementary/Alternative Medicine and Biomedicine in Clinical Settings* (Unpublished PhD dissertation, University of Toronto).

Hollenberg, D., Cowan, B. and Wright, N. (2007), 'New and Uncharted Territory: Integrative Medicine in the Medical Radiation Sciences', *The Canadian Journal of Medical Radiation Technology* Winter 2007, 23–30.

Hui, K., Hui, E.K. and Johnston, M.F. (2006), 'The Potential of a Person-centered Approach in Caring for Patients with Cancer: A Perspective from the UCLA Center for East-West Medicine', *Integrative Cancer Therapies* 5:1, 56–62.

Kleinman, A. (1988), *The Illness Narrative* (New York: Basic Books).

Mann, D., Gaylord, S. and Norton, S. (2004), 'Moving Toward Integrative Care: Rationales, Models, and Steps for Conventional-care Providers', *Complementary Health Practice Review* 9, 155–72.

Nutbeam, D. (1998), 'Health Promotion Glossary', *Health Promotion International* 13:4, 349–64.

Saks, M. (1994), *Professions and the Public Interest: Medical Power, Altruism and Alternative Medicine* (London: Routledge).

Shuval, J.T., Mizrachi, N. and Smetannikov, E. (2002), 'Entering the Well-guarded Fortress: Alternative Practitioners in Hospital Settings', *Social Science & Medicine* 55, 1745–55.

Williams, A. (1998), 'Therapeutic Landscapes in Holistic Medicine', *Social Science & Medicine* 46:9, 1193–1203.

World Health Organization (2002), *Traditional Medicine Strategy 2002–2005* (Geneva, Switzerland: Author).


**Internet-based References**

Atun, R. (2004), *What are the Advantages and Disadvantages of Restructuring a Health Care System to be more Focused on Primary Care Services?* Electronic Document. Available at http://www.euro.who.int/document/e82997.pdf.

Croasdale, M. (2006), *Allopathic Med School Enrollment Rises 2.2%: Private and Public Schools Boost Class Sizes in Response to Physician Shortage Concerns.* Electronic document. Available at http://www.ama-assn.org/amednews/site/free/prsc1113.htm.

Hollenberg, D. (2007), 'How do Private CAM Therapies Affect Integrative Health Care Settings in a Publicly Funded Health Care System?', *Journal of Complementary and Integrative Medicine* 4:1, Article 5. Electronic document. Available at http://www.bepress.com/jcim/vol4/iss1/5.

World Health Organization (1978). *Declaration of Alma-Ata*. Electronic document. http://www.euro.who.int/AboutWHO/Policy/20010827_1.

# PART 3
# Places and Settings

Chapter 11

# Considering the Clinic Environment: Implications for Practice and Primary Health Care

Valorie A. Crooks and Gina Agarwal

Common to all health services is that care is always delivered *somewhere* (i.e., in or from a particular place). By this we do not only mean in a locality or community, something that is emphasized in the Alma-Ata Declaration (WHO 1978) in relation to primary health care (PHC), but also, at the micro-scale, in a specific space. These spaces are often clinical in nature such as a family doctor's office, a community health clinic, or possibly even a hospital depending on the medical care being delivered. Those which are not clinical in nature, such as a care recipient's home, may be transformed for the purpose of care delivery, either temporarily or permanently (Yantzi et al. 2007). In this chapter we focus specifically on those clinical spaces within which care informed by the tenants of PHC is delivered, and particularly on the community-based health clinic.

Geographers have demonstrated how built, symbolic, and natural environments can contribute to the development and maintenance of health (Gesler and Kearns 2002; Kearns and Gesler 1998; Williams 1999). Spaces within which care is delivered are often designed to promote healing (Kearns and Barnett 1997, 1999, 2000). Within this research, particular attention has been paid to how the physical layout, colours, and furnishings within these sites contribute to the overall therapeutic environment (Arneill and Devlin 2002; Gesler 1999). Other messages may, however, compete with establishing an environment conducive to healing. For example, consumerist messages and designs are becoming increasingly common in health care settings (Kearns and Barnett 2000; Kearns et al. 2003) and may disrupt the therapeutic or healing atmosphere with messages of consumption (Crooks and Evans 2007). Furthermore, the existence of equipment – such as charts, scales, carts, and drip stands – adds to the formal feel of medical places, possibly making them seem unwelcoming and possibly less therapeutic (Andrews and Shaw 2008; Mahon-Daly and Andrews 2002). Another important variable that adds to the complexity of these settings and the environments they create for those within them is that of power. Clinical spaces are imbued with power (Gesler et al. 2004; Gillespie 2002) and are sites where power relations are played out (Halford and Leonard 2003; Poland et al. 2005; Rapport et al. 2006).

Although the inquiries discussed above have been wide-ranging, few studies to date have considered the specific meaning of clinical space (see Andrews and Shaw 2008). Responding to this gap in the literature, in this chapter we explicitly consider the role of the clinic space in providing PHC. Our intent is to consider the multiple and complex ways in which space impacts upon individuals' abilities to both deliver and receive care. We have three specific objectives: (1) to determine those specific elements of the community-based health clinic that may positively or negatively affect the development of a therapeutic/healing environment; (2) to consider their implications for clinical practice; and (3) to explicate the role of the clinic(al) space in implementing (i.e., putting in place) the principles of PHC outlined in the Alma-Ata Declaration. In order to address these objectives we first introduce a new, previously unpublished, case study that serves to establish the relationship (if not affect) between specific elements of the clinic environment and patients' development of a therapeutic relationship with their practitioners. Because this is a new case study we do provide some background about research design and context prior to sharing findings. We then move to discuss the findings of this and other studies in light of the aforementioned objectives.

## Case Study

The development of a 'therapeutic relationship' between doctor and patient is both special and, many would argue, needed (Balint 1972; Andrews and Evans 2008). In the case of women who are experiencing clinical depression, having a relationship of mutual trust and respect between doctor and patient is important. In this case study we asked 20 women experiencing clinical depression to talk about how they conceptualize and develop a therapeutic relationship with their family doctors. While what they shared has important implications for clinical practice, the findings also reveal the ways in which the clinic's environment also significantly shaped their experiences of the (potentially therapeutic) doctor-patient relationship.

*Context*

Clinical depression is a chronic condition that leads to feelings of sadness and/or despair and changes in physiological functioning (e.g., to sleep, appetite, libido) which can have significant implications for people's everyday functioning. It can be a disabling disorder and requires both diagnosis and often ongoing treatment and/or management. At times it is difficult to diagnose, sometimes leading to misdiagnosis or a lack of any diagnosis at all (Farvolden et al. 2003). Clinical depression is also a highly gendered disorder for a variety of reasons – including that men and women have exposure to different kinds of social stressors – and is more prevalent among women (Kessing 2005; Wilhelm et al. 2002). Further, certain women are more vulnerable to onset, including: single mothers (Cairney et al. 1999), pregnant women (Gotlib et al. 1989), and those with a history of physical abuse (MacMillian et al. 2001) and/or childhood adversity (Davies et al. 1997).

The development of a supportive relationship and good communication style between doctor and patient is important in all circumstances (Roter 2000). This is particularly the case for the treatment and management of women's clinical depression. In the case of treating mental disorders specifically, a 'therapeutic relationship' between doctor and patient is often desired whereby change may be effected (Andrews and Evans 2008; Castonguay and Constantino 2006). However, issues such as the: minimizing or disregarding of women's reporting of symptoms (DiCowen 2003), particularly those which are invisible (Crooks 2006); hierarchical and gendered nature of the health care system (Sebrant 1999); and traditionally (but not necessarily contemporarily) patriarchal nature of the doctor-patient communication style (Goodyear-Smith and Buetow 2001) can challenge the development of a supportive and potentially therapeutic relationship. Thus, it is important for us to look to women experiencing clinical depression themselves to tell us about what qualities they desire in a family doctor and how they establish the parameters of what is and is not appropriate to discuss, as is done in this case study.

*Study Design*

The purpose of this study was to determine what characteristics clinically depressed women viewed as being desirable in their relationships with family doctors. We also wanted to understand how women drew boundaries in this relationship, including in determining what kinds of information to share with their doctors. Our goal was to greater understand the potential for development of a therapeutic relationship between clinically depressed women and their family doctors. Women who had been treated for clinical depression within a six month period and were patients at a specific family practice clinic (located in London, England) were invited to participate in an interview.

Interviews were undertaken at a location of the participant's choosing which in some cases was the woman's house, and in others the doctor's clinic.[1] A semi-structured interview schedule guided the conversation. The guide was organized into four main sections: (1) background and demographic information; (2) experiences at the clinic; (3) the doctor-patient relationship; and (4) experiences of clinical depression.

Thematic analysis of the dataset was employed. Our first step was to randomly select two transcripts from the dataset to independently review in order to develop a preliminary coding scheme. We then held a meeting to devise the scheme. In order to enhance consistency, one investigator then undertook the first stage of

---

1   Ethical consent for the study was granted by the Imperial College School of Medicine and informed consent was obtained prior to the start of the interview.

coding. A second investigator reviewed the coded transcripts and made adjustments where necessary thereby undertaking the second stage of coding. Thematic issues emerging from the dataset after coding were then reviewed by both of us.

*Key Finding – Clinic Environment*

The women spoke openly about the characteristics they desired in a family doctor. Those they viewed as important included having family doctors who are: in a certain age range, available for appointments, able to schedule adequate amounts of time for appointments, female, good listeners, and willing to give referrals to specialists. An unanticipated finding emerging from the interviews was that many of the women talked about the importance of the clinic environment in shaping their interactions with doctors. These comments were unprompted, in that none of the questions in the interview guide probed issues of the clinic environment. These comments suggest to us that the potential for establishing a therapeutic relationship is shaped in part by the space within which care is provided. Specifically, women commented about features such as: the location of the practice, how busy/crowded the clinic was, elements of the décor and ambiance, the layout, and the presence and use of technology.

The number of other people present in the clinic space was a key concern of the women, including in relation to the overall patient load and also the number of clinicians and staff available. The presence or absence of others greatly influenced how they read and interpreted the clinic environment. For example, one participant noted:

> ... the feeling in the practice when you arrive, busy ... exhausted receptionist, people fed up, waiting, a feeling of dilapidation and stress ... You can hear people being put off on the phone and you could hear 'no no I can't put you through to the doctor now' 'no no you'll have to call back' and that makes you feel worse because you don't want to call back again at an inappropriate time.

In this instance the busyness of the clinic and also the audible sounds of people being turned away affected how she felt about the clinic and her ease within the space. The comments of another woman were similar: 'And of course I feel guilty about all the other people waiting in the waiting room.' Again in this instance the visual cue of seeing waiting patients in the clinic led to a certain amount of unease while there. Another participant indicated that the fullness of the waiting room would, at least in part, determine how much time she intended to take up when interacting with the doctor so as to ensure that other patients would be seen.

While several of the women were concerned with how many other staff and patients were in the clinic during their visits, there was a specific concern regarding this issue that some shared. Specifically, a busy clinic environment may threaten the privacy that they desired both as patients in general and individuals experiencing clinical depression. One woman had this to say about the clinic:

… in here [present clinic], if I could walk in here, you've got the reception desk. You've got people passing. You can hear other people's conversations – which I don't particularly want to hear … There [former clinic] you just walked in, you saw the receptionist, they knew you had an appointment, then you went straight into the waiting room. So you couldn't hear what the other person coming after you, about their business.

For the most part the women desired an environment that reflected personalized care and maintained privacy. The presence or absence of others in the clinic space was, however, not the only indicator of whether or not privacy would or could be maintained. It was clear from the women's comments that the maintenance of privacy was also linked explicitly to clinic layout and specifically to the presence of curtains and dividing walls in the clinical practice space (i.e., the practice room/ area) itself.

The presence or absence of technology, and computers specifically, was another element of the clinic environment that the women commented about. The presence of a computer within the clinical practice space, for example, was considered a deterrent to establishing a good, and potentially therapeutic, doctor-patient relationship by a number of women. These women's comments convey this sense:

- It is amazing. Okay, press a button, you've got all the notes in front of you which saves time. That's fine, impersonal … That's why the approach now, I think, between the patient and doctor is slowly disappearing. There is contact but from a distance.
- I do find these days that they're so busy looking at their computer screens and trying to read your notes and things they're not ever looking at you or … I sometimes feel that they're not listening to you, um, just cos they're – I'm sure they are, but they're – you know, they're reading your notes on the computer and, and things.
- See that's the other thing, the old doctor didn't have a computer, did she? So that's why it was necessary to know somebody because they knew their history, but now they have a computer and well that's the trouble they sit there for the first five minutes going through the computer to see who you are.

It is clear from these comments that the presence of a computer in the clinical practice space may hamper the establishment of a therapeutic relationship between doctor and patient as it can impede the flow of conversation or necessitate 'contact … from a distance' as one participant put it. At the same time, it was understood by many that the use of computing technologies in the clinic in general was indicative of modern medical practice and therefore generally desirable.

The design, layout, and overall ambience of the clinic environment were important to the women and were raised in their discussions about developing a therapeutic relationship with their doctors. Interestingly, a participant directly related the appearance of the clinic to those who practiced within it and the kind of care they delivered, noting:

> It's quite modern and ... the actual presence of the surgery [clinic] does make a difference to your perception of, of what the doctors are going to be like and the area was nice. And, I don't know, my first impressions were encouraging ... You know, they obviously put a lot of time and effort into doing it up and sort of makes you think that they'll, they'll put a bit of time and effort into other things.

From this woman's comments and others it was obvious that the overall look of the clinic was important in relation to how they understood their doctor and built a (potentially therapeutic) relationship with him/her. The presence of material items such as magazines and plants also contributed to creating an environment that was read as 'very comfortable and ... cozy' by the women.

Finally, one of the things that most often prompted the women to discuss the clinic environment was when they were attempting to make comparisons between clinics. Two types of comparisons were made. The first, and most common, occurred when the women spoke about their experiences as a patient in the present clinic versus one they may have frequented in the same or another urban community in the past. Differences between past and present clinics were often explained in relation to both the clinic environment and also the quality of care received. A less common comparison came from those women who at one time had been seen in a rural clinic. In these instances the bases for comparison tended to be on urban versus rural practice differences in general and less on specific elements of the clinic environment such as layout and design. For example, one woman noted that 'Well, I, you know, everything's very civilized in the country, you've always got much nicer surgeries [clinics] and, and, um, doctors have got more, more space and stuff.' In general, experiences as a patient at multiple clinics enhanced the women's abilities to critically reflect on the role of the clinic environment in shaping the potentially therapeutic doctor-patient relationship.

**Implications for Clinical Practice**

In the above case study, the women's comments clearly indicate how the clinic environment could impact upon their levels of comfort and possibly interactions with doctors. Because of this we can imagine that there are implications for information sharing specifically. Another study of ours in which we interviewed family caregivers and individuals with chronic illnesses identified this same issue; specifically, it was revealed that interviewees had concern that the use of computers during the doctor-patient interaction led to more impersonal care (Crooks and Agarwal 2008; see also Risdale and Hudd 1994). The presence of computers within the clinic practice space in particular may disrupt the conversational flow between doctor and patient, potentially disrupting information sharing and possibly the establishment of informational continuity of care. Informational continuity of care refers to the 'extent to which information about past care is used to make current care appropriate to the patient' (Haggerty et al. 2007, 340) and thus involves both information sharing and record keeping. If a patient does not feel comfortable sharing information then not only may a therapeutic relationship never be developed but informational continuity may never be achieved. An implication for clinical practice, then, is that doctors and other clinic staff must be sensitive as to how and when computers are used during an appointment as they are an important part of the clinical environment.

The above case study serves to remind us that there are, in fact, multiple micro-spaces within the clinic that serve different purposes. Many of the women's comments singled out the waiting room. Sutherland (2002) contends that this space offers a first impression of the clinic. Further, as one of the women noted, it also communicates information about those who practice in the clinic (Arneill and Devlin 2002). Waiting rooms may serve as a site of socialization (Kearns 1991) and often include one or more of: comfortable furnishings, artwork, magazines, play areas for children, and educational posters and pamphlets (Arneill and Devlin 2002; McGrath and Tempier 2003; Sutherland 2002). In contrast, the clinical practice space serves a very different purpose and also has a much different environment. Rapport et al. (2007) distinguish between the doctor's work area and the patient examination area within this space. In a study of family doctors' workspaces they found that doctors treated the examination table as a sacred site and even those with cluttered workspaces made sure this area was not particularly messy. Items typically found in this space include examination tables, files and file folders, book shelves, clocks, desks, wall postings such as posters and artwork (Rapport et al. 2006, 2007). In considering the implications for clinical practice, perhaps the most important one is that if the environment of a particular micro-space (e.g., waiting room, clinical practice space) is not in keeping with its purpose then care delivery and receipt could potentially be compromised.

Building on our previous point, not only do different micro-spaces within the clinic serve different purposes, but the same space may also be read differently by different occupants. Such an understanding relies on the notion that health care spaces are texts that can be read and interpreted (Kearns et al. 2003). Rapport et al.'s (2006, 2007) study of family doctor workspaces reminds us that these doctors will read the clinic space from the perspective of a worker and clinician. This is a positioning and social location that is quite different from that of the patient, a non-clinical staff member, or most others who enter the clinic. In the case study reported above, the busy-ness that some women commented on as being distracting and sometimes unwelcome may simultaneously be read by the doctor as a sign of a thriving practice; these are two very different readings of the same environment. Further, the women's reading of such an environment as distracting may be informed, at least somewhat, by the clinical depression making them more sensitive to their surroundings; meanwhile, another patient or patient group may interpret this element of the clinic environment quite differently.

Gillespie (2002) conducted a case study of a family planning clinic in order to specifically examine how architecture can inform and is informed by the cultural practices associated with delivering health care. In undertaking observations in the clinic she noted that architecture, including the physical separation between doctors and patients, enforced different uses of the clinic space and ultimately the ways in which different individuals read and interpreted its purpose. An important finding of her work is that clinic architecture reflects and enforces power hierarchies between different users. This finding is confirmed in Rapport et al.'s (2006) study of family doctor workspaces whereby they contend that power relations are manifest in the clinic space. Examples include that doctors typically have larger and more comfortable chairs to sit in than do patients in the clinical practice space and also that there are struggles over control and use of space between junior and senior doctors practicing in the same clinic. We can understand that power is central to how the clinic environment is read and understood by different occupants. Crooks (2006) contends that factors such as how the health care system is organized and operated, the social construction of professional and lay knowledge(s), and people's performances as patients are all roots of the power that is expressed in this space. An implication for clinical practice is that the different readings of the clinic space must be acknowledged (see Andrews and Evans 2008) and attempts to reduce the ways in which the environment enforces power hierarchies of all sorts may need to be made. This would be consistent with the current push in health care to move away from paternalistic and towards patient-centred models of care.

Perhaps the biggest implication of what we have discussed above for clinical practice is that best practice may not be achieved if the clinic environment is disruptive. In our case study we contend that the establishment of a therapeutic relationship between doctor and patient may be compromised if the clinic environment is interpreted in certain ways (e.g., as a space that does not allow for privacy to be maintained). In Rapport et al.'s examination of the family doctor workspace they found that some doctors practiced in what they referred to as 'unsuitable space' (Rapport et al. 2006). Such space was inappropriate for their needs and they assert that it may negatively impact upon clinical practice. Oppositely, others worked in spaces they characterized as 'supportive' in that they actually enhance the doctor's ability to achieve best practice. In their work they establish a clear link between the clinic environment, including the organization and arrangement of the clinical practice space, and practice outcomes. Finally, Gillespie (2002), in her case study of a family planning clinic, questions how a highly feminized, ableist, and hetero-normative clinic interior can be welcoming for groups such as gay men and lesbian and disabled women. If the clinic environment is not inclusive then best practice may further be threatened by the exclusion of certain groups in need of care. Because of this it is important that the setting in which care is delivered, in this case the environment of the community-based clinic, be considered in best practice guidelines.

**Implications for Implementing Primary Health Care**

As outlined in the Alma-Ata Declaration of 1978, PHC is clearly care that is delivered in the community, for the community, and even by community members. Explicit references to 'community' are numerous in the principles of the Declaration itself. The Declaration conveys a clear sense of the local, but not necessarily the specific location or site of care delivery. We must recognize, however, that care informed by these principles of PHC is delivered specifically from somewhere in the community, and often the clinic. Perhaps one of the most relevant implications of this understanding for putting in place the kind of PCH envisioned in the Alma-Ata Declaration is to consider if and how the community should inform the clinic environment. The Declaration makes clear that care should reflect the socio-politico-economic characteristics of the country and community where it is being delivered. With this in mind, should the clinic environment be reflective of the community? There may not even be a single answer to this question, but our discussion in the previous section of the chapter certainly helps to inform how we might approach beginning to find answers.

Ensuring that care is delivered accessibly and equitably to citizens is central to achieving the kind of PHC outlined in the Alma-Ata Declaration. In relation to our focus on the clinic environment, we must ask: can the clinic environment either enhance or detract from achieving accessible and equitable PHC? Central to this is understanding how to balance the needs/desires of one patient or group against those of others. In our own case study, for example, the depressed women's need for the clinic environment to be not overly busy and also reflective of privacy may not be shared by others, including those practicing in the clinic. Whose needs should be prioritized in such an instance? A related issue is that we know very little about those who do not visit particular clinics due specifically to their environments. There could be many bases for such an outcome, some of which are due to choice (e.g., deciding to not go to a clinic because its environment is not reflective of the standard of care one would want to receive) and others to need (e.g., being prevented from using a clinic due to physical barriers or it not being a scent-free environment). If this indeed does happen then the accessibility and equitability of care being provided is challenged because the clinic environment prohibits meeting certain needs and facilitating care receipt by specific community members. It also impacts upon the delivery of care which is not 'first contact' to these individuals in that if a primary care clinic or some other setting of PHC delivery has an environment not conducive to care receipt for certain individuals then referrals to specialists may never be made thereby further negatively affecting access to care.

An implication of our entire discussion up to this point is that doctors must be cognizant of the role that the clinic(al) environment plays in shaping their clinical practice and also determining how individuals interpret the care they receive. This point illustrates the complexity of implementing the principles of PHC outlined in the Alma-Ata Declaration as planners, clinicians, and health service administrators alike must be in tune not only with the goal of achieving health for all but also of creating welcoming and even health-promoting clinic spaces. In recognizing this we do, however, acknowledge that there are many constraints that impact upon developing a health-promoting, healing, therapeutic, or even welcoming clinic space. Lack of funding, for example, places limitations on elements such as design and architecture and even the implementation of technologies within clinics. If those designing these spaces do not have knowledge about the types of relationships we have discussed above between place, health, and health care, then the potential role of the clinic environment in creating a therapeutic experience may never be fully considered. Further to this, there is a need to balance desires regarding efficiency and patient flow in designing the clinic space and its environment with features such as the overall ambiance and aesthetic. Factors such as these will play a significant role in determining how much focus, if any, is placed on achieving the desired therapeutic clinic environment.

Adding to our last point, we also recognize that the restructuring of health care delivery, and of primary care specifically, has implications for the kinds of environments that are present in modern clinics. The primary care systems of several developed nations are encouraging the formation of large, integrated health care teams that include practitioners from multiple disciplines. An outcome of this is that these teams will need complex clinical spaces, ones not necessarily in keeping with traditional community-based primary care clinics. Based on what we have discussed above, there are, of course, many implications of this including that: different parties will read these clinic environments differently, and certain disciplinary practices may necessitate the absence or presence of features of the clinic environment that are found to be incongruent the practices of others. We need to be attentive to if and how system restructuring and other new health care initiatives inform the clinic environments in which care, including that informed by the principles of PHC, is delivered and received.

With the increasing desire to use evidence to inform health care decision-making, a final implication of what we have discussed in this chapter for putting in place the principles of PHC in care delivery is to consider whether or not we need to formally evaluate the role of the clinic environment in directing things such as clinical practice, the doctor-patient relationship, and ultimately health/therapeutic outcomes. Doing so would, admittedly, be quite challenging for many reasons. Alternatively, we might consider somehow including the clinic environment in our evaluation of PHC initiatives more generally. Haggerty et al. (2007) used a Delphi consultation process with Canadian PHC experts to achieve consensus in determining and defining attributes of the care process central to PHC. They identified five attributes, they are:

1. clinical practice attributes (e.g., accessibility, comprehensiveness);
2. structural dimensions (e.g., information management, multidisciplinary practice);
3. person-oriented dimensions (e.g., respectfulness, cultural sensitivity);
4. community-oriented dimensions (e.g., equity, community participation); and
5. system performance dimensions (e.g., accountability, efficiency).

A purpose of their identification of these attributes is to assist in evaluating and measuring PHC initiatives through developing operational definitions. Given this, perhaps a starting point is to determine if and where the clinic environment has a place within these various attributes or if it is a dimension of its own. This is just one approach; there are of course numerous others that could be adopted in order to evaluate the role of the clinic environment in PHC provision. The larger question that must be addressed is whether or not this is needed or even wanted.

### Conclusion

At the outset of this chapter we set for ourselves three objectives, the first of which was to determine those specific elements of the community-based health clinic that may positively or negatively affect the development of a therapeutic/healing environment. In order to address this we presented a new case study focused on the findings of 20 interviews conducted with women experiencing clinical depression. It was found that characteristics such as the number of people in the clinic, whether or not the layout was conducive to maintaining privacy, the presence of computers in the clinical practice space, and the overall ambiance and design of the clinic were important to the women and shaped how they interpreted the kind of care delivered in the clinic, the doctor-patient relationship they had established, and ultimately the therapeutic potential of this relationship. Perhaps most interesting is that the women's comments about these characteristics of the clinic environment were unprompted. It must be recognized that this listing of characteristics of the clinic environment that may affect healing or the therapeutic outcomes of care is specific to our case study and thus not exhaustive. In spite of this, the case study does highlight a number of important relationships between place, health care, and healing.

Our second objective was to consider the implications of the clinic environment and its importance for clinical practice. We identified four specific implications, the first of which pertains to information sharing and establishing informational continuity of care. Specifically, doctors and other clinic staff must be sensitive as to how and when technologies in the clinic environment are used during an appointment in order to avoid disrupting the flow of information in the doctor-patient interaction. The second implication references our contention that the clinic is made up of micro-spaces that serve different purposes. Given this, if the environment of each micro-space is not in keeping with its purpose then care delivery and receipt may be compromised. Thirdly, the same space in the clinic may be read differently by different people. An implication of this is that attempts must be made to reduce the ways in which the clinic environment enforces power hierarchies, including those which may be expressed during clinical practice. The fourth and final implication we identified is that best practice may not be achieved if the clinic environment is disruptive. These implications are significant, and there are certainly others that could be identified. It is thus essential that those in clinical practice be open to considering the multiple ways in which the clinic environment shapes outcomes such as the quality of their care delivery and the interactions they have with patients.

Finally, our third objective was to explicate the role of the clinical space in implementing the principles of PHC outlined in the Alma-Ata Declaration. To do this we discussed the implications that clinic environments hold for how we can put such care provision into place. We identified five implications, specifically that we need to consider: (1) if and how the community (e.g., local socio-economic status, cultural norms) should inform the clinic environment; (2) how the clinic environment may enhance or detract from achieving accessible and equitable care; (3) that doctors must be cognizant of the role that the clinic(al) environment plays in shaping clinical practice and how patients receive their care; (4) how the restructuring of health services may have outcomes for the clinic environment; and (5) whether or not, and how, the clinic environment should be considered when evaluating PHC initiatives. These implications are certainly far ranging and are relevant to a number of the principles of PHC identified in the Alma-Ata Declaration. Importantly, each of these points also serves as a research agenda in that they are all issues we need to know more about in order to more effectively deliver PHC, including in the clinic setting.

### Acknowledgements

A Canadian Institutes of Health Research (CIHR) Interdisciplinary Capacity Enhancement sub-grant (through the Health Care, Technology and Place CIHR Strategic Training Institute) provided some funding for us to work collaboratively on this study.

### References

Andrews, G.J. and Shaw, D. (forthcoming), 'Clinical Geography: Nursing Practice and the (Re)making of Institutional Space', *Journal of Nursing Management*.
Andrews, G.J. and Evans, J. (forthcoming), 'Understanding the Reproduction of Health Care: Towards Geographies in Health Care Work', *Progress in Human Geography*.

Arneill, A.B. and Devlin, A.S. (2002), 'Perceived Quality of Care: The Influence of the Waiting Room Environment', *Journal of Environmental Psychology* 22, 345–60.

Balint, M. (1972), *The Doctor, His Patient and the Illness* (New York: International Universities Press).

Cairney, J., Thorpe, C., Rietschlin, J. and Avison, W.R. (1999), '12-month Prevalence of Depression Among Single and Married Mothers in the 1994 National Population Health Survey', *Canadian Journal of Public Health* 90:5, 320–4.

Castonguay, L.G. and Constantino, M.J. (2006), 'The Working Alliance: Where are We and Where Should We Go?', *Psychotherapy: Theory, Research, Practice, Training* 43:3, 271–9.

Crooks, V.A. (2006), '"I Go on the Internet; I Always, You Know, Check to See What's New": Chronically Ill Women's Use of Online Health Information to Shape and Inform Doctor-patient Interactions in the Space of Care Provision', *ACME: An International E-Journal for Critical Geographies* 5:1, 50–69.

Crooks, V.A. and Agarwal, G. (2008), 'Chronically Ill Patients' and Family Caregivers' Perspectives on the Use of Computer Technologies in Family Practice: Implications for the Development of Informational Continuity of Care', in J.B. Garner and T.C. Christiansen (eds), *Social Sciences in Health Care and Medicine* (New York: Nova Science Publishers).

Crooks, V.A. and Evans, J.D. (2007) 'The Writing's On the Wall: Decoding the Interior Space of the Hospital Waiting Room', in A. Williams (ed.), *Therapeutic Landscapes* (Aldershot and Vermont: Ashgate Publishing), 165–180.

Davies L., Avison W.R. and McAlpine D. (1997), 'Significant Life Experiences and Depression Among Single and Married Mothers', *Journal of Marriage and the Family* 59:2, 294–308.

DiCowden, M.A. (2003) 'The Call of the Wild Woman: Models of Healing', *Women & Therapy* 26:3–4, 297–310.

Farvolden, P., McBride, C., Bagby, R. and Ravitz, M. (2003), 'A Web-based Screening Instrument for Depression and Anxiety Disorders in Primary Care', *Journal of Medical Internet Research* 5:3, e23.

Gesler, W.M. (1999) 'Words in Wards: Language, Health and Place', *Health & Place* 5, 13–25.

Gesler, W., Bell, M. and Curtis, S. et al. (2004), 'Therapy by Design: Evaluating the UK Hospital Building Program', *Health & Place* 10, 117–128.

Gesler, W. and Kearns, R. (2002) *Culture/Place/Health* (New York: Routledge).

Gillespie, R. (2002), 'Architecture and Power: A Family Planning Clinic as Case Study', *Health & Place* 8, 211–220.

Goodyear-Smith, F. and Buetow, S. (2001), 'Power Issues in the Doctor-patient Relationship', *Health Care Analysis* 9, 449–462.

Gotlib, I.H., Wiffen, V.C. and Mount, J.H. et al. (1989), 'Prevalence Rates and Demographic Characteristics Associated with Depression in Pregnancy and the Postpartum', *Journal of Consulting and Clinical Psychology* 57:2, 269–74.

Haggerty, J., Burge, F. and Levesque, J.-F. et al. (2007), 'Operational Definitions of Attributes of Primary Health Care: Consensus Among Canadian Experts', *Annals of Family Medicine* 5, 336–44.

Halford, S. and Leonard, P. (2003), 'Space and Place in the Construction and Performance of Gendered Nursing Identities', *Journal of Advanced Nursing* 42:2, 201–8.

Kearns, R.A. (1991), 'The Place of Health in the Health of Place: The Case of the Hokianga Special Medical Area', *Social Science & Medicine* 33:4, 519–30.

Kearns, R.A. and Barnett, J.R. (1997), 'Consumerist Ideology and Symbolic Landscapes of Private Medicine', *Health & Place* 3, 171–80.

Kearns, R.A. and Barnett, J.R. (1999), 'To Boldly Go? Place, Metaphor and the Marketing of Auckland's Starship Hospital', *Environment and Planning D: Society and Space* 17, 201–26.

Kearns, R.A. and Barnett, J.R. (2000), '"Happy Meals" in the Starship Enterprise: Interpreting a Moral Geography of Health Care Consumption', *Health & Place* 6, 81–93.

Kearns, R.A., Barnett, J.R. and Newman, D. (2003), 'Placing Private Healthcare: Reading Ascot Hospital in the Landscape of Contemporary Auckland', *Social Science & Medicine* 56, 2303–15.

Kearns, R.A. and Gesler, W. (1998), *Putting Health into Place: Landscape, Identity and Well-being* (Syracuse: Syracuse University Press).

Kessing L.V. (2005), 'Gender Differences in Patients Presenting with a Single Depressive Episode According to ICD-10', *Social Psychiatry and Psychiatric Epidemiology* 40, 197–201.

MacMillan, H.L., Fleming, J.E. and Streiner, D.L. et al. (2001), 'Childhood Abuse and Lifetime Psychopathology in a Community Sample', *American Journal of Psychiatry* 158:11, 1878–83.

Mahon-Daly, P. and Andrews, G.J. (2002), 'Liminality and Breastfeeding: Women Negotiating Space and Two Bodies', *Health & Place* 8:2, 61–76.

McGrath, B.M. and Tempier, R.T. (2003), 'Is the Waiting Room a Classroom? (Letter to the Editor)', *Psychiatric Services* 54:7, 1043.

Poland, B., Lehoux, P., Holmes, D. and Andrews, G.J. (2005), 'How Place Matters: Unpacking Technology and Power in Health and Social Care', *Health and Social Care in the Community* 13:2, 170–180.

Rapport, F., Doel, M.A., Greaves, D. and Elwyn, G. (2006), 'From Manila to Monitor: Biographies of General Practitioner Workspaces', *Health: An Interdisciplinary Journal for the Social Study of Health, Illness and Medicine* 10:2, 233–51.

Rapport, F., Doel, M.A. and Elwyn, G. (2007), 'Snapshots and Snippets: General Practitioners' Reflections on Professional Space', *Health & Place* 13:2, 532–44.

Ridsdale, L. and Hudd, S. (1994), 'Computers in the Consultation: The Patient's View', *British Journal of General Practice* 44, 367–9.

Roter, D. (2000), 'The Enduring and Evolving Nature of the Patient-physician Relationship', *Patient Education and Counseling* 39, 5–15.

Sebrant, U. (1999) 'Being Female in a Health Care Hierarchy: On the Social Construction of Gender and Leader Identity in a Work Organization Having a Predominance of Women', *Scandinavian Journal of Caring Sciences* 13:3, 153–8.

Sutherland, M. (2002), 'Rating Waiting', *BMJ* 324, 1443.

WHO (World Health Organization) (1978), *The Alma-Ata Declaration* (Geneva: World Health Organization).

Wilhelm, K., Roy, K. and Mitchell, P. et al. (2002), 'Gender Differences in Depression Risk and Coping Factors in a Clinical Sample', *Acta Psychiatrica Scandinavia* 106, 45–53.

Williams, A. (ed.) (1999), *Therapeutic Landscapes: The Dynamic Between Place and Wellness* (Lanham: University Press of America).

Yantzi, N.M., Rosenberg, M.W. and McKeever, P. (2007), 'Getting Out of the House: The Challenges Mothers Face When their Children have Long-term Care Needs', *Health and Social Care in the Community* 15:1, 45–55.

Chapter 12

# Within and Beyond Clinics: Primary Health Care and Community Participation

Robin Kearns and Pat Neuwelt

In this chapter we examine the significance of spaces of provision in the delivery of primary health care (PHC) alongside the notion of *community participation*. We draw primarily from our experience in New Zealand and focus initially on the clinic, identifying this site as a necessary but not sufficient component of primary care. We then consider *practice* – both as an organizational entity and changing practises of medicine – in the context of greater emphasis on community participation within international PHC care policy imperatives.

By way of a departure point, a foundational experience for us was PHC in the Hokianga locality of northern New Zealand in the late 1980s. In this district, formerly designated a Special Medical Area (in which doctors were salaried by central government to care for under-serviced populations) we observed that clinics were serving as *de facto* community centres. This was clearly a radically different dynamic to the usual clinical setting, in which medical power prevails and people are at least implicitly constructed as patients if not diseased bodies (Philo 2000). While bodies were still examined behind closed doors, it was nonetheless apparent that these rural clinics constituted an integral component of the landscape of community cohesion and information exchange. People were known to the professionals and were observed visiting the often rudimentary facilities to not only see a doctor, but also to interact with people waiting for medical attention. Indeed, on one memorable occasion the first author's observational research in a clinic waiting room was interrupted by a *kuia* (elderly Maori woman) entering, greeting those present, and then kissing everyone, including the researcher. Here, the medical service was clearly a necessary but not sufficient component of a primary care system based on community and relationships (Kearns 1991). Later, in the early 1990s, resistance to amalgamation of this small-scale and community-focussed rural health system into a regional entity could be partly explained by the degree of stake and participation in health care experienced by local people (Kearns 1998).

In Auckland, 300 km to the south (2006 population 1.25 million), we observed a contrasting urban model of primary care develop in the 1990s. Here, environments rather than interactions were apparently valued by patients. New accident and medical (A&M) clinics, often sited near Auckland's shopping malls, featured flashy interiors and comforting surrounds including televisions and free coffee. These branded clinics offered after-hours medical care for routine general practitioner (GP) consultation and minor accidents, employing shift-working doctors. This situation minimized the likelihood that one would encounter any continuity of care (Barnett and Kearns 1996). Little wonder, then, that when over 200 patients were surveyed at two sites, aspects of the clinic ambience were more commonly remarked upon by patients than cost or continuity of care (Kearns and Barnett 1997).

The foregoing anecdotes highlight two different constructions of people's affiliation with the sites of primary care in New Zealand. In the former example, people are *participants* in that they appear to feel sufficiently at home at community clinics to converse with others and even drop in and pass the time of day. The clinics are more than locations or sites; they are meaningful places (Kearns 1993). Indeed, recent developments in the Hokianga district have seen the link between place and participation develop to the point that community members have become central to the governance of their local health system. In the contrasting latter example, material rather than relational aspects of the setting appeared to hold most appeal to users who demonstrated no particular stake in the clinic and are thus better regarded as consumers of, rather than participants in, primary care.

We contend that conventional GP practices in New Zealand, which are the main setting of primary care delivery, lie somewhere between these poles of places of participation and sites of consumption. GP practices are generally constrained in their engagement with, and contribution to, local places by their focus on one-on-one doctor-patient relations. While the doctor(s) and other staff may be well-known within a neighbourhood, patients-as-consumers do not necessarily know each other, thus constraining the possibility of a medical practice contributing to community development.

Further points of distinction between the two types of primary care settings are their business model of care and their underlying ideology. The Hokianga health service, now Hokianga Health Enterprise Trust (HHET), is community owned and governed. Despite the emergence of primary health organizations (PHOs) over the past decade in New Zealand, the majority of general practices remain small businesses (or medium-sized, in the case of the A&M clinics), owned by GPs (with additional corporate stakeholders in the case of A&M clinics). In Hokianga, care is free at point of contact while in general practices, there are patient co-payments. The ideology in each setting also differs. In the former, the basis of care is teamwork between GPs, primary care nurses and health promoters. The service delivered is comprehensive PHC, which includes illness care, preventive care and broad health-promoting initiatives. In keeping with comprehensive PHC, it also involves community participation with a community development agenda. In contrast, GP practices deliver medical care with some preventive care, including screening, immunization and individual health education, and involve no community participation.

In the remainder of this chapter, we examine the distinction between participants in, and consumers of, primary care and ask how place is implicated in this distinction. We note that there are many concepts related to community participation in the literature, such as community involvement (Kahssay and Oakley 1999), public involvement (Department of Health 2001), community development (Raeburn 1998), community empowerment (Laverack and Labonte 2000; Laverack and Wallerstein 2001), community capacity (Goodman et al. 1998) and community competence (Eng and Parker 1994). We use the term *participation* in this chapter in an inclusive sense, incorporating the broad concepts of community engagement and public involvement, with the goals of empowerment and development. Our goal is to connect participation and place. The rationale for our chapter is that there is a push in many Western countries to incorporate consumer and community input more widely in the health sector. We examine community participation as rhetoric and reality, and conclude that while many primary care practices are constructed as *catchments,* current policy imperatives – at least in New Zealand – seek a closer engagement with place, endorsing the status of the clinic as a potential catalyst for community development.

## Context

In anchoring our case study in New Zealand, we recognize the need at the outset to establish some context. Although it is one of the more urbanized countries in the Western world, New Zealand nonetheless has a significant rural and scattered population whose access to services is often constrained by a dissected coastline and rugged topography, as well as distance from urban centres. Maori comprise 14 percent of the national population, a significant proportion of which live in rural areas (especially in the north and east of the North Island). Maori and migrants from the Pacific Islands comprise two of the more disadvantaged groups in terms of health, with Maori experiencing approximately ten years lower life expectancy than non-Maori. Unlike Maori who generally have connections to rural communities, Pacific People who comprise approximately seven percent of the national population, are mainly resident in larger urban centres and especially Auckland. Auckland also includes the only significant population of migrants from Africa, Asia and the Middle East whose diverse health needs are recognized as poorly met by established PHC services (Lawrence and Kearns 2005).

The Hokianga example is illustrative of a third sector primary care service (non-profit, non-government) which, in New Zealand, developed in the 1980s in response to the inaccessibility of primary care to low income groups, due to cost and other barriers (Crampton et al. 2004a). In rural areas, many such services developed were *kaupapa Maori* (established, owned, and run by Maori) (Crengle et al. 2004a; Crengle et al. 2004b). Despite evidence that these services deliver care to populations with high needs, they only serve about ten percent of the New Zealand population (Crampton et al. 2004b). Significant inequalities in access to primary care have remained in New Zealand, particularly for Maori (Crengle et al. 2004a) and people of Pacific island nations (Young 2000). The implementation of the 2001 Primary Health Care Strategy (King 2001) in New Zealand has been an attempt to reduce those inequalities in access, with the aim of achieving more equitable health outcomes. The Strategy called for the establishment of not-for-profit PHOs which were to be groups of providers offering illness care and health promotion to enrolled populations and which had community participation in their governance.

**Community Participation in PHC**

In this section, we review the philosophical foundations of participation in primary care, asking more fundamentally what it is and how might it connect with ideas of place. In so doing we elaborate on the core contention of our chapter: that participation in health care by non-professionals in the planning and development of primary care services is commonly purported to improve access to, and the appropriateness of, services. However, although participation may arguably improve access, there is only limited research evidence that establishes participation as improving patient outcomes. As a result, attitudes about the relative importance of community participation tend to be polarized along ideological lines. This polarization implicitly centres on understandings of community. Restrictive understandings of community tend to see membership as reflected in service use (e.g. as occurred in the Community Health Centre movement in the United States). A natural – and limiting – extension of this view is the use of a consumerist model in which health *service* improvement – rather than health improvement – is the ultimate goal (Callaghan and Wistow 2002).

*Origins*

Community participation in health service planning and development formally entered the policy arena with the Alma-Ata Declaration on PHC in 1978. This first international conference on PHC held at Alma-Ata, USSR (now Kazakhstan) and sponsored by the World Health Organization (WHO) and the United Nations Children's Fund (UNICEF), defined PHC in the following way (WHO and UNICEF 1978, Section VI):

> Primary health care is essential health care based on practical, scientifically sound and acceptable methods and technology made universally accessible to individuals and families in the community *through their full participation* and at a cost that the community can afford to maintain at every stage of their development *in the spirit of self-reliance and self-determination* (emphasis added).

This same definition was adopted for the New Zealand 2001 PHC Strategy (King 2001, 29).

The Declaration was a radical statement which added a socio-political perspective to the definition of health and to the purpose of health services. It highlighted the importance of community participation in health care planning and delivery by redefining health as a human rights issue and participation in health planning as a democratic principle. The intent to shift control from health professionals and bureaucrats to communities was clearly implied in the tone and content of the Declaration: 'The people have the right and duty to participate individually and collectively in the planning and implementation of their health care' (WHO and UNICEF 1978, Section IV).

Although the Declaration has been quoted widely in health literature, the application of its principles internationally has been patchy and slow. A quarter of a century later, one of the PHC principles highlighted in the Declaration has made its way into New Zealand health policy, with the requirement that PHOs involve *iwi* (Maori tribes) and communities in their governance. There is, however, significant confusion over the interpretation of the policy. In part this lack of clarity stems from participation, like community, being a concept which defies a single interpretation, despite a vast literature in health and the social sciences. It has been argued that participation has become an umbrella term for a people-centred approach to development, and its interpretations reflect the ideological position of those initiating the participatory process (Kahssay and Oakley 1999).

*Approaches*

There are two approaches to public participation with contrasting agendas, defined in the literature as democratic and consumerist. Participation as a democratic principle defines citizens as having rights to access public services and duties to participate collectively in society (Lupton et al. 1998). The concepts in the Alma-Ata Declaration were based on this notion of participatory democracy; yet, much of the health literature has focused on consumer use of services, underplaying the place of community involvement in health care. The two are inextricably linked, as one (consumption of services) at least potentially informs the other (experience of participation).

In distinguishing between consumer and community involvement in health care planning, a representation of the relationship between consumers and citizens is worthy of consideration. Draper (1997) conceptualized a continuum, from the individual receiving care through to the collective of citizens. The distinction between citizens and consumers has been argued as being central to claims about legitimate involvement in (rather than simply receipt of) health care (Callaghan and Wistow 2002). At different levels of the continuum from citizen to consumer, the democratic (or citizenship) and consumerist approaches to public participation with contrasting agendas have been identified (Croft and Beresford 1992; Lupton et al. 1998; Pickard and Smith 2001). Each of these approaches is argued to have different interests (Lenaghan 1999) and, some have argued, different value sets (Mooney and Blackwell 2004). These will be considered next.

The consumerist approach to participation (consumerism) is a service-led approach, which highlights consumer involvement for the purposes of identifying individual preference and enhancing market competitiveness (Brown 2000). The focus of this approach is on the rights of consumers to information, access, choice and redress in relation to a specific service or product (Lupton et al. 1998). In the consumerist approach, health care is a product for which the market takes priority over active political engagement between members of society (Cawston and Barbour 2003). Cawston and Barbour (2003) suggested that patients have been re-defined as *consumers* of health services as part of the resurgence of neo-liberalism in the latter twentieth century. With this redefinition came a shift in emphasis from community and the wider determinants of health to individual priorities and interactions with health services, even if expressed through consumer organizations. Client or consumer satisfaction has been the main outcome measure of consumerism in health, which feeds into quality improvement programmes, yet its validity has been increasingly questioned (Sitzia 1997, cited in Cawston and Barbour 2003).

There is a large literature on consumer preferences in primary care, but in this section we will highlight only systematic literature reviews. Wensing et al. (1996), for instance, derived a list of key aspects of primary care that Dutch consumers considered to be important. These aspects included availability, accessibility, organization, cooperation between providers, effective and competent medical care, doctor-patient relationship, information, counselling and support. Further work was undertaken on how consumers have ranked these aspects by doing a systematic review of the literature. In the overall rating of factors, for the 57 international studies reviewed, 'humaneness, competence/accuracy and patient involvement in decisions' were ranked as the three top priorities by consumers of primary care (Wensing et al. 1998).

A subsequent systematic literature analysis of consumer characteristics as predictors of PHC preferences applied the ranking identified by Wensing et al. (1998) and found that the most significant predictors of preferences were age and economic status. Elsewhere, Jung et al. (2003) reviewed 145 quantitative studies of the association between patient characteristics and PHC preferences, and found evidence of a divide in people's value orientation towards health care, with patients either seeing themselves as active participants or as passive consumers.

Significantly, perhaps, the age, education and health status of those seeing themselves as active participants reflects a position arguably less deferential to medical power and, less encumbered by morbidity, in a stronger position to be assertive. However, in order to interpret the notion of passive primary care consumers, we argue for the importance of considering the social context. Rogers and colleagues (1999) adopted a social process approach to understanding help-seeking behaviour in primary care, and concluded that consumer demand for health care is created and shaped by not only personal factors, such as past experiences of health care, but also by system-level factors. Using the example of after-hours primary medical care, they argued that there has been a failure to link supply with demand. Out of hours care used to be considered a mainstay of primary care; however changes to arrangements for out of hours care have tended to be made without consumer consultation.

Elsewhere, a review of research on patient involvement in health service planning revealed that such involvement does lead to service improvements, including improved physical access, better provision of information for consumers, and extended opening hours (Crawford et al. 2002). Crawford and colleagues highlighted clear evidence that the collective involvement of patients can lead to more accessible health services.

*Critiques*

A number of authors have provided a critique of the consumerist approach to participation in health. Cawston and Barbour (2000) contended that it is a reductionist approach to participation, based on a medical view of disease in which health is a product for consumption, and information is interpreted by doctors, pharmaceutical companies and government agencies. Consumerism has also been criticized as favouring the wealthy and articulate, with no consideration of the non-consumer, and no ability to cope with conflicting consumer views (Lenaghan 1999). Rifkin and colleagues (2000) have argued that the usual exclusion of non-users from participation means that those most vulnerable have no voice in health planning.

It has been argued that members of the public have a dual relationship with those commissioning and providing health care services; that of services users and of citizens. Lenaghan (1999, 48) distinguishes between the two perspectives:

> For example, an individual patient has an immediate and personal interest in the particular service which they receive. Citizens, meanwhile, have a broader and longer-term interest in the health service, as voters, taxpayers and members of the community; they are interested in what happens not only to themselves, but also to their families, neighbours and fellow citizens, both now and in the future. These two perspectives must always be taken into account, but policy makers need to be clear about which perspective is appropriate to which type of health care decision, and be aware of the potential for conflict [between the two].

There is evidence from the United Kingdom that primary care organizations overwhelmingly view service quality improvement as the rationale for involving users and the public (Milewa et al. 2002). Callaghan and Wistow (2002) have argued that the role for citizen involvement is often much less clear in policy than is the role for consumer involvement, with regard to service improvement. Yet, it is this wider concept of community or citizen involvement that our present chapter aims to address. Our next task is to uncover some of the motives underlying the pursuit of a more participatory approach to PHC.

*Agendas*

Participatory democracy in PHC can be distinguished from consumer participation by the breadth and modes of participation (Jewkes and Murcott 1998). The coordinator of Health Care Aotearoa, a national network of third sector PHC organizations in New Zealand, clarified the distinction in regard to PHOs:

> There seems to be confusion between 'consumer' and 'community'. The priority must be on the community. Naturally, there should be regular surveying and dialogue with the actual patients who make up the register of the PHO [primary health organisation]. But that is different to discussion with the community—this discussion will be driven by the community, on their venues, following their protocols, and led by their agenda. The patient (or consumer) work will be much more driven by the PHO (which will) want to ensure that the services they offer are appropriate and effective (Glensor 2001, n.p.).

In other words, participation most fundamentally involves engagement with not just the collective of users or consumers, but with the community which supports and encircles a service. As such, whereas it might be assumed that the place to engage the consumer is that location where service is consumed (i.e., the clinic), the place of the community may or may not also be a site of clinical significance. In fact, as Glensor suggests above, engaging with communities must take place beyond the clinic in a community venue. Thus, a range of gathering places may be appropriate in which to engage with a community (e.g., a community hall, a church, a school). The appropriate place will, to an extent, be determined by whether the community in question is based on propinquity (in which case there may only be one logical meeting place, particularly in rural areas where others may have closed with population decline) or is a community-of-interest (in New Zealand, a *marae*, or traditional community gathering place, is often the most appropriate site at which to engage with Maori; on account of high levels of religious affiliation, a church hall would often be the most appropriate place to meet with a Pacific community).

The foregoing views resonate with a nuanced view of place that sees on-the-ground sites of significance linked recursively with identity through people's 'place-in the world' (Eyles 1985; Kearns 1993). In other words, meeting about health concerns at *marae* will affirm Maori views of health because that venue is where Maori identity is secure and reproduced through gathering on a regular basis. Indeed, it may make most sense to not only discuss health care services on a *marae* but actually bring health care itself to such venues and away from clinical settings. This shift is occurring in the Tairawhiti area where Turanga Health's kaumatua (elders) programme regularly brings together up to 200 older people, rotating between rural *marae* each fortnight for a day of sociability, informative talks and heath checks (e.g., podiatry, blood pressure). In this way community participation in health is being enhanced by a shift in the place of health care into a familiar, non-clinical and, hence, less intimidating environment.

## Actualising Community Participation in NZ

How can participation in PHC best be actualized? In the literature on participation in health and health care, Arnstein's (1969) conceptualization of participation as a ladder has been adapted from its origins in the planning literature (Neuwelt 2007). Following the ladder metaphor, the absence of participation is sited at the bottom and the highest level of participation, total community control, on the top rung of the ladder. Neuwelt and Crampton (2005) argue that the impact of involving consumers and communities in primary care planning is dependent on the level of their involvement in the organization, commenting that consumer and community participation, in at least the New Zealand health sector, has often been nominal. In their estimation, the provision of feedback through information-sharing or community consultation does not necessarily have any impact on service planning.

The agenda underpinning a policy of community involvement in primary care will determine the working definition of community and the model of community involvement. Similarly, the approach taken in defining relevant communities in the PHC setting will define the model of community participation and be directly linked to the aims of such participation. These three factors are closely interrelated, and can be argued to be the underlying determinants of community participation in PHC. For example, if communities are defined as consumers, then participation will follow a consumerist model. If communities are defined as citizens, then the model will be of citizen accountability; however, if consumer and citizen involvement is defined by need, on the basis of health inequalities, then the model of participation has the potential to be an empowering one. An empowering approach to community participation is one that, through the involvement of members of disadvantaged communities, leads to positive social change or community development. A key facet of positive social change is improved access to care.

### Access as Impetus

While access to primary care is regarded as a key 'problem', it is an issue that has been conceived in largely spatialized and financial terms within policy discourses. In other words, accessibility has generally been seen as meeting challenges of overcoming barriers of distance and cost. The example of A&M clinics that we opened with addresses another dimension of access: hours of opening. Their longer hours of opening allow uninsured patients to trade off cost against availability and in so doing, the pressure of patient numbers is removed from public hospital emergency departments. However, is getting to a clinic enough to satisfy the issue of access? To address this question we need to consider aspects of what Joseph and

Phillips (1984) called *effective accessibility* that implicate the issues of place and participation central to this book. In other words, a range of patient, provider and community setting characteristics may impact upon access above and beyond the spatial facts of relative distance or proximity.

By way of background, New Zealand health care consumers have traditionally accessed the health sector through GP services. GPs have maintained the right to charge co-payments, and have operated principally within a small business model of practice. This situation has arguably encouraged the construction of patients as *consumers*. In terms of resistance to this implicit construction, the growth of third sector primary care organizations, such as the union health centres in the late 1980s and ethnic-specific health services in the 1990s, was a direct response to the unmet needs of groups of people who were disadvantaged by this model (Crampton 1999). As discussed earlier, Maori providers and, more recently, Pacific providers have become part of the landscape of third sector primary care delivery in New Zealand. Third sector organizations, established using a community governance model, which have nearly always featured health professionals working on salaries, have had a positive impact on the utilization of services by high needs groups and on selected outcomes, while promoting the rights of communities to participate in health and health care decision-making (Crampton et al. 2001). Third sector primary care organizations have been found to serve a higher proportion of low income families than their for-profit GP counterparts (Crampton et al. 2004). Their consumers are younger, largely non-European and present with more health problems than consumers in private general practices. Further, they have tended to adopt the PHC model of practice, in keeping with the Alma-Ata Declaration. These differences highlight the importance for communities, policy makers and purchasers of the community-governed non-profit model in improving access to care.

*Community Characteristics*

Community characteristics have a bearing on how participation is best actualized. The 1995 *Hauora: Maori Standards of Health III* report (Pomare et al. 1995) highlighted the fact that Maori have difficulty accessing health care services, and it called not only for support for Maori organizations but also for mainstream services to ensure that they are affordable, accessible, appropriate and acceptable for Maori. Maori health providers have been established over the past 20 years by iwi, hapu or pan-tribal groups to provide more culturally appropriate health services to Maori. A survey of the medical and nursing care delivered by Maori primary health providers indicated that they generally serve Maori with high health needs, and that they demonstrate responsiveness to those needs (Crengle et al. 2004b). These findings suggest that through the heightened participation of Maori in the provision of care, providers are in turn increasing access to care for those who live in areas of high relative deprivation. The 2001 New Zealand PHC Strategy now requires PHOs to be responsive to the needs of local Maori.

Given this set of links between culture, place and health, what patient, provider and community setting characteristics might affect access above and beyond issues of physical distance or proximity? By way of example, in a qualitative study investigating Maori views about health and their experiences of health care, Cram et al. (2003) reported that participants' experience and knowledge of Pakeha (European) doctors was far from positive. Maori had found that 'persistence and assertiveness were required, often in the face of cultural misunderstandings, if good health care was to be obtained from existing systems' (Cram et al. 2003, 7). Maori also described *whakamaa* (reticence or shyness) as a potential barrier to health care. This was seen to potentially limit Maori from seeking health care or from telling a doctor what was wrong with them. Rapport, involving GPs taking time to put patients at ease, was described as vital to the doctor-patient relationship and could counteract *whakamaa*. The researchers reported that traditional healing played a role for many Maori, particularly in terms of *rongoa* (traditional healing practices) and *wairua* (the spiritual realm). Some participants in the study reported integrating both 'Maori and Pakeha medicines' (i.e., both traditional healing and Western medicine).

Similar findings are reported from Pacific peoples. A qualitative study of Pacific consumers of health care in New Zealand included 175 focus group participants of Samoan, Tongan, Cook Islands or Niuean ethnicity, and highlighted some important issues for Pacific peoples (Pacific Health Research Centre 2003). All ethnic groups described a desire for access to ethnic-specific health services. For all four groups, the major barriers to health care services were reported to be cost, communication and language difficulties, lack of transport and waiting times to see a doctor. As with Maori, traditional healing was reported to play an important role for many Pacific peoples, across all the ethnic groups studied, and was often used in conjunction with Western medicine.

Research into the experiences of new migrants/settlers at an Auckland PHC clinic reveals some similar access issues to those of Maori and Pacific peoples. Lawrence and Kearns (2005) found that resettlement issues, particularly relating to uncertainty in immigration status, employment and housing, were reported as having a significant influence on people's health. The major barriers to accessing health services for new settlers were described as cost, language, cultural differences and physical access issues. In summary, a range of studies have identified issues of cultural nature including language and communication as barriers to effectively accessing and participating in contemporary PHC in New Zealand.

*Linking Participation and Access*

Many rationales underlie the involvement of consumers and communities in the planning and delivery of health care, though in health policy these rationales are often implicit, at best (Dwyer 1989; Kahssay and Oakley 1999). Arguments for participation, broadly based on either democratic or consumerist principles, include to: legitimate policy; increase patient/community compliance; improve the appropriateness of services and of policy; redistribute power and resources; and offer less tangible benefits to communities, such as an increased local knowledge base. In the foregoing section, we have reviewed recent New Zealand health policy experience, highlighting ways in which some of the foregoing arguments necessarily implicate place more than others. For example, redistribution of power and resources and increasing local knowledge potentially enhance people's 'place-in-the-world' and, if more than lip-service is to be achieved, require taking health concerns into community settings.

Arguably, the question which reveals the true purpose of any process for involving communities in the public sector in general, and health care in particular, is 'who benefits?' (Dwyer 1989). A further and related rationale for community participation is its potential to reduce inequalities in health (National Health Committee 1998). Health equity was an explicit goal of PHC in the Alma-Ata Declaration. A more fundamental justification for community participation in primary care in New Zealand is found in the Treaty of Waitangi, which defined political participation as a right of the Treaty partners, Maori as *tangata whenua* (people of the land) and the Crown. The Treaty gives Maori the unequivocal right to self-determination and the right to access culturally appropriate services. In light of this fact, it is critical that PHOs involve *tangata whenua* in their participatory processes.

In reality, however, many participatory processes benefit the health sector more than communities. Some critics argue that community involvement in the public sector is a clear example of the neoliberal political agenda of many Western governments to shift responsibility from public servants to communities (Petersen and Lupton 1996). From this perspective, governments are potentially the greatest beneficiaries of community involvement in the public sector. While there is undoubtedly some truth to that view, there is research evidence that community participation in primary care can and does benefit patients, communities and health services.

Here, and in illustration, we can return to one of our opening examples, the HHET, in the far north of New Zealand. In the Hokianga district, since the late 1940s, community clinics have ensured that medical care is taken to the people in remote settings and in ways that are sufficiently acceptable that the communities claim the clinics as *de facto* community centres. In the current era of PHOs as an outgrowth of the New Zealand Primary Care Health Strategy, elements of the Hokianga health system that have been in place since the 1980s such as a high level of community ownership (e.g., elected trustees from each community served by a clinic) are now being promoted as the 'gold standard' in the early twenty-first century. To this extent, the leadership in participation has come from a system of primary care paradoxically set up as a 'special medical area' precisely because access was a problem. Now developments in a place once deemed peripheral are central to the discourse of primary care in New Zealand.

By way of anecdote, the HHET coordinated a project to improve the quality of drinking water at *marae*, following water sampling tests that revealed high levels of contamination. The project involved the installation of water treatment systems at 36 *marae* throughout the Hokianga region between 2000 and 2003 (Wigglesworth pers. comm. 2003). Decisions about the selection of water treatment systems and building plans were made together with the *marae* committees, and local people were employed to do the work (Jellie 2003). The project was deemed by all involved to have been a huge success and bears echoes of the early and foundational principles of PHC and ongoing concerns for the WHO (Phillips 1990). The high level of community involvement in, and control of, the water project was considered to have been the key to this success (Stuart et al. 2003).

We are therefore arguing that ensuring effective and ongoing participation within PHC is actually an important aspect of improving access to health itself. In so doing, an expanded view of the significance of place arises when a 'sum is greater than the parts' perspective is adopted and there is an acknowledgement of the limits of a perspective that places the doctor at the centre of the health equation. This view is reflected in urban as well as rural initiatives. In Auckland, for instance, one PHO recognized the potential to engage the burgeoning Muslim migrant community through offering culturally appropriate exercise opportunities. A public swimming pool was contracted to restrict entry at a certain time each week to women only and this allowed Muslim women to take swimming lessons while suitably attired (Lawrence 2006).

**Conclusion**

In this chapter, we have reviewed the philosophical and practical bases for participation in primary care, surveying international literature as well as drawing on New Zealand-based research examples. We have distinguished between involving consumers and communities, in that the former constitute actual and current patients whereas the latter include all who have a stake in a service – even if only potential rather than actual users. Engaging communities, we contend, involves broadening the construction of place in primary care, for a clinic is a necessary but not sufficient component of a community-focused PHC system. Rather, a range of other significant sites including schools, halls, churches and even shopping malls could, hypothetically, comprise significant places for health development and delivery. Importantly, while our examples have been anchored in New Zealand, the issues we have discussed are relevant to the challenges facing other nations. We note, however, that our use of New Zealand examples is more than convenience, given that this nation's engagement with PHC principles within national policy is arguably among the more advanced in the western world.

It is ironic that, despite community participation being at the heart of the original PHC model, it has largely been applied within the practice of health promotion, rather than PHC settings. As we have demonstrated, the tide is turning on this trend in New Zealand. The requirement for PHOs in New Zealand to involve community and *iwi* (Maori tribal) participation in their governing processes is challenging. Yet the very act of engaging in participatory processes is arguably complementing the underlying goal of reducing health disparities in New Zealand through improved access to appropriate PHC.

While there is relatively little research evidence directly relating to the benefits and costs of participation, there are many examples in New Zealand and elsewhere of the benefits of community involvement in health. Third sector experience has shown that community participation is not simply an establishment issue, but is an ongoing process of partnership in decision-making between providers, *iwi* and communities.

Demonstrating that participation may involve engagement on the community's terms and in community places begs the question of whether community participation is a process or an outcome, a means or an end in itself (Rifkin 1996; Kahssay and Oakley 1999; Laverack and Wallerstein 2001). As a means, the aim of community participation is to encourage people's cooperation with externally introduced health services or projects. As a process, participation itself is the goal. It is a process of empowerment, an instrument of positive social change for disadvantaged individuals and groups. Experts in the field (Kahssay and Oakley 1999; Rifkin 1996) now argue that participation in PHC needs to be both a means and an end, to have maximum mutual benefit for those individuals and groups, as well as for services.

In summary, geographers have overlooked the diverse roles and workings of clinics in the past. Our attention to participation as a key practice of PHC suggests that while the efficacy of biomedical power remains pervasive in institutional settings, smaller clinics imbued with the imperatives of PHC hold potential for a more multi-dimensional empowerment through community participation. In terms of thinking about clinics as places, we conclude that participation as reality (rather than rhetoric) holds capacity to diffuse the medical power and the largely unidirectional 'gaze' that has conventionally been associated with clinical settings (Philo 2000). In some of the foregoing examples, clinics are either *de facto* community centres, or, conversely, community settings that can take on clinical functions. Such observations arguably dilute the social construction of the clinic as a potentially oppressive environment and instead suggest we should regard the clinical setting inflected with PHC principles as something embedded in, rather than set apart from, community dynamics.

We conclude that community participation can affirm fundamental aspects and rights associated with citizenship. Further, its pursuit and practice through shifting primary medical care towards the more holistic model of PHC may be a very important step towards the reduction of health inequalities. Our advocacy for the adoption of a broad notion of community, along with an empowerment approach to participation, has been in keeping with an interpretation of the concept of community participation linked to the concept of community development. In many ways, the starting point for this work was the Declaration of Alma-Ata, and its notion of comprehensive PHC, which had built on the successes of early PHC pioneers. New Zealand has a unique history of community participation in health, some of which has been acknowledged in this chapter. Three decades on from Alma-Ata, as New Zealand moves towards making the principles of community participation in PHC a reality, investigating the positive impacts on both services and the health of the communities they serve remains a research priority.

## References

Arnstein, S.R. (1969), 'Ladder of Citizen Participation', *Journal of the American Institute of Planners*, 35, 216–23.

Barnett, J.R. and Kearns, R.A. (1996), 'Shopping Around? Consumerism and the Use of Private Accident and Medical Clinics in Auckland, New Zealand', *Environment and Planning A*, 28, 1053–1075.

Brown, I. (2000), 'Involving the Public in General Practice in an Urban District: Levels and Type of Activity and Perceptions of Obstacles', *Health and Social Care in the Community*, 8:4, 251–59.

Callaghan, G. and Wistow, G. (2002), *Public and Patient Participation in Primary Care Groups: New Beginnings for Old Power Structures* (Leeds: Nuffield Institute for Health).

Cawston, P.G. and Barbour, R.S. (2003), 'Clients or Citizens? Some Considerations for Primary Care Organisations', *British Journal of General Practice* 53, 716–22.

Cram, F., Smith, L. and Johnstone, W (2003), 'Mapping the Themes of Maori Talk About Health', *New Zealand Medical Journal* 116:1170, 357–63.

Crampton, P. (1999), *Third Sector Primary Health Care: A Report Prepared for the National Health Committee* (Wellington: Department of Public Health, Wellington School of Medicine).

Crampton, P., Davis, P. and Lay-Yee, R. et al. (2004a), 'Comparison of Private For-profit With Private Community-governed Not-for-profit Primary Care Services in New Zealand', *Journal of Health Services Research & Policy* 9 (Suppl 2:S2), 17–22.

Crampton, P., Lay-Yee, R. and Davis, P. (2004b), *The National Primary Medical Care Survey (NatMedCa): 2001/02 Report 2: Primary Care in Community-governed Non-profits: The Work of Doctors and Nurses* (Wellington: Ministry of Health).

Crampton, P., Woodward, A. and Dowell, A. (2001), 'The Role of the Third Sector in Providing Primary Care Services: Theoretical and Policy Issues', *Social Policy Journal of New Zealand* 17, 1–21.

Crawford, M., Rutter, D. and Manley, C. et al. (2002), 'Systematic Review of Involving Patients in the Planning and Development of Health Care', *British Medical Journal* 325, 1263–65.

Crengle, S., Crampton, P. and Woodward, A. (2004a), 'Maori in Aotearoa/New Zealand', in J. Healy and M. McKee (eds), *Accessing Health Care: Responding to Diversity* (Oxford: Oxford University Press).

Crengle, S., Lay-Yee, R. and Davis, P. (2004b), *Maori Providers: Primary Health Care Delivered by Doctors and Nurses: The National Primary Medical Care Survey (NatMedCa): 2001/02 Report 3* (Wellington: Ministry of Health).

Croft, S. and Beresford, P. (1992), 'The Politics of Participation', *Critical Social Policy* 29, 20–44.

Department of Health (2001), *Shifting the Balance of Power Within the NHS: Securing Delivery* (London: Department of Health).

Draper, M. (1997), *Involving Consumers in Improving Hospital Care: Lessons from Australian Hospitals* (Canberra: Commonwealth Department of Health and Family Services).

Dwyer, J. (1989), 'The Politics of Participation', *Community Health Studies* XIII:1, 59–64.

Eng, E. and Parker, E. (1994), 'Measuring Community Competence in the Mississippi Delta: The Interface between Program Evaluation and Empowerment', *Health Education Quarterly* 21:2, 199–220.

Eyles, J. (1985), *Senses of Place* (Warrington: Silverbrook Press).

Glensor, P. (2001), *Health Care Aotearoa Submission on "Minimum Requirements and Best Practice Guidelines for Priimary Health Organisations"* (Wellington: Ministry of Health HCA).

Goodman, R.M., Speers, M.A. and McLeroy, K. et al. (1998), 'Identifying and Defining the Dimensions of Community Capacity to Provide a Basis for Measurement', *Health Education and Behavior* 25:3, 258–78.

Jellie, M. (2003), *Kaupapa Maori Drinking Water Quality and Community Development*, Public Health Association of New Zealand Conference 2003: Tino 7 Rangatiratanga in Public Health, Turangawaewae Marae, Ngaruawahia.

Jewkes, R. and Murcott A. (1998), 'Community Representatives: Representing The "Community"?', *Social Science & Medicine* 46:7, 843–58.

Joseph, A.E. and Phillips, D.R. (1984), *Accessibility and Utilization: Geographical Perspectives on Health Care Delivery* (New York: Harper & Row).

Jung, H.P., Baerveldt, C. and Olesen, F. et al. (2003), 'Patient Characteristics as Predictors of Primary Health Care Preferences: A Systematic Literature Analysis', *Health Expectations* 6, 160–181.

Kahssay, H.M. and Oakley, P. (eds) (1999), *Community Involvement in Health Development: A Review of the Concept and Practice*, Public Health in Action (Geneva: World Health Organization).

Kearns, R.A. (1991), 'The Place of Health in the Health of Place: The Case of the Holkianga Special Medical Area', *Social Science and Medicine* 33, 519–530.

Kearns, R.A. (1993), 'Place and Health: Towards a Reformed Medical Geography', *The Professional Geographer* 45, 139–147.

Kearns, R.A. (1998), 'Going it Alone: Community Resistance to Health Reforms in Hokianga, New Zealand', in R.A. Kearns and W.M. Gesler (eds), *Putting Health into Place: Landscape, Identity and Well-being* (Syracuse: Syracuse University Press), 226–247.

Kearns, R.A. and Barnett, J.R. (1997), 'Consumerist Ideology and the Symbolic Landscapes of Private Medicine', *Health & Place* 3, 171–180.

King, H.A. (2001), *The Primary Health Care Strategy* (Wellington: Ministry of Health), 30.

Laverack, G. and Labonte, R. (2000), 'A Planning Framework for Community Empowerment Goals Within Health Promotion', *Health Policy & Planning* 15:3, 255–62.

Laverack, G. and Wallerstein, N. (2001), 'Measuring Community Empowerment: A Fresh Look at Organizational Domains', *Health Promotion International* 16:2, 179–85.

Lawrence, J. (2006), *Evaluation of Asian Women's Swimming Programme* (Auckland: Procare Ltd.).

Lawrence, J. and Kearns, R.A. (2005), 'Exploring the "Fit" Between People and Providers: Refugee Health Needs and Health Care Services in Mt Roskill, New Zealand', *Health & Social Care in the Community* 13, 451–61.

Lenaghan, J. (1999), 'Involving the Public in Rationing Decisions. The Experience of Citizens Juries', *Health Policy* 49:1–2, 45–61.

Lupton, C., Peckham, S. and Taylor, P. (1998), 'Understanding Public Involvement', in C. Ham and C. Heginbotham (eds), *Managing Public Involvement in Healthcare Purchasing* (Buckingham: Open University Press), 44–61.

Milewa, T., Harrison, S. and Ahmad, W. et al. (2002), 'Citizens' Participation in Primary Healthcare Planning: Innovative Citizenship Practice in Empirical Perspective', *Critical Public Health* 12:1, 39–53.

Mooney, G.H. and Blackwell, S.H. (2004), 'Whose Health Service is it Anyway? Community Values in Healthcare', *Medical Journal of Australia* 180, 76–8.

National Health Committee (1998), *The Social, Cultural and Economic Determinants of Health in New Zealand: Action to Improve Health* (Wellington: National Advisory Committee on Health and Disability), 3–108.

Neuwelt, P. (2007), *Community Participation Toolkit: A Resource for Primary Health Organisations* (Wellington: Steele Roberts).

Neuwelt, P.M. and Crampton, P. (2005), 'Community Participation in Primary Health Care in Aotearoa New Zealand', in K. Dew and P. Davis (eds), *Health and Society in Aotearoa New Zealand*, 2nd edition (Auckland: Oxford University Press), 194–210.

Pacific Health Research Centre (2003), *Pacific Island Primary Health Care Utilisation Study* (Auckland: Pacific Health Research Centre).

Petersen, A. and Lupton, D. (1996), 'The Duty to Participate', *The New Public Health: Health and Self in the Age of Risk* (St Leonards: Allen & Unwin), 146–73.

Phillips, D.R. (1990), *Health & Health Care in the Third World* (New York: Longman).

Philo, C. (2000), 'The Birth of the Clinic: An Unknown Work of Medical Geography', *Area* 32, 11–19.

Pickard, S. and Smith, K. (2001), 'A "Third Way" for Lay Involvement: What Evidence so Far?', *Health Expectations* 4, 170–79.

Pomare, E, Keefe-Ormsby, V., Ormsby, C., Pearce, N., Reid, P. Robson, B. and Watene-Haydon, N. (1995), *Hauora: Maori Standards of Health III: A Study of the Years 1970–1991* (Wellington: Te Ropu Rangahau Hauora a Eru Pomare, Huia Publishers).

Raeburn, J. and Rootman, I. (1998), *Community Development. People-centred Health Promotion* (Chichester: John Wiley & Sons Ltd.).

Rifkin, S.B. (1996), 'Paradigms Lost: Toward a New Understanding of Community Participation in Health Programmes', *Acta Tropica* 61, 79–92.

Rifkin, S.B., Lewando-Hundt, G. and Draper, A.K. (2000), *Participatory Approaches in Health Promotion and Health Planning: A Literature Review* (London: Health Development Agency).

Rogers, A. Hassell, K. and Nicolaas, G. (1999), *Demanding Patients? Analysing the Use of Primary Care* (Buckingham: Open University Press).

Stuart, K., Foote, J. and Jellie, M. et al. (2003), 'Learning from the Stories of Nga Puna Wai o Hokianga', *PHA News* VI, 1.

Wensing, M. and Grol, R. (1996), 'Indicators of the Quality of General Practice Care of Patients with Chronic Illness: A Step Towards the Real Involvement of Patients in the Assessment of the Quality of Care', *Quality in Health Care* 5, 73–80.

Wensing, M. and Jung, H.P. et al. (1998), 'A Systematic Review of the Literature on Patient Priorities for General Practice Care. Part 1 Description of the Research Domain', *Social Science and Medicine* 47, 1573–1588.

WHO and UNICEF (1978), *Declaration of Alma-Ata*, International Conference on Primary Health Care, Alma Ata, USSR, WHO/Europe.

Young, N. (2000), 'Pacificans' Access to Primary Health Care in New Zealand: A Review', *Pacific Health Dialogue* 4:2, 68–74.

# Chapter 13

# Providers of Care in the Home: Sustainable Partners in Primary Health Care?

Nicole M. Yantzi and Mark W. Skinner

> [Primary health] care … shifts and changes within and across settings with implications for who cares and the form that care takes (Milligan et al. 2007, 136).

Literature on health policy, research and practice has examined, identified and critiqued the increasing importance of the *home* as an essential place of care and caregiving (see Dyck et al. 2005). Yet, it is curious that in policies and debates regarding primary health care (PHC) in Western health care systems, the essential role of providers of care in the home is often ignored. Such is the case, despite the fact that PHC practices like diagnosis, treatment, prevention and health promotion already take place in the home as part of the growing reliance on home and community care provisioning (see Cooper et al. 2003). Indeed, changes in health care policy, service delivery and technology have blurred the boundaries between types of health care work such as ongoing chronic care and management and first contact PHC, and between types of care providers in the home such as paid and unpaid. The importance of homes and households has, however, not gone entirely unnoticed within the health and social sciences, and key studies in Europe, North America and elsewhere are calling for more attention to providers of care in the home within the continuum of PHC (see Buetow 2004; Haggerty et al. 2003).

In this chapter, we contribute to the literature arguing for inclusiveness within the continuum of PHC by examining the crucial yet under-valued role of providers of care in the home, conceived of here as anyone involved in managing, coordinating and delivering in-home health care whether it is paid or unpaid, formal or informal, public or private. Specifically, we draw on two qualitative studies examining the provision of care in the home and community in Ontario, Canada, which taken together feature 84 in-depth interviews with directors, case managers, paid staff, volunteers and family members involved in caring for children and older adults (Skinner 2005; Yantzi 2005). Our overall purpose is not to distinguish or differentiate among this diverse group but to illustrate from a provider perspective how these individuals are *de facto* partners in meeting the PHC needs of people living with chronic conditions. We contend that their inclusion within PHC policy is incomplete without first understanding their vulnerability within the formal health care system not to mention the attendant paradox of society's growing reliance on these under-resourced and, in the longer-tem, potentially unsustainable partners in PHC.

Providers of care in the home are in jeopardy of not fulfilling their roles. They are being charged with greater responsibility for caregiving due to devolution, divestment and downloading in the health system. At the same time, they are asked to do so in the face of the increasing number and complexity of needs for care in the home and the socio-economic and demographic pressure on individuals, families and communities to cope independently (Skinner and Joseph 2007). We contend that appreciating the vulnerability and sustainability of care provisioning in the home is a prerequisite for its successful inclusion within the continuum of PHC and for development of informed PHC policy.

The chapter unfolds in three major parts. We begin with a review of the literature that conceptualizes the home as an important place where PHC occurs and critiques the neglect of in-home care provisioning within prevailing PHC policy. Set within the context of PHC in Ontario, we then re-analyse existing interviews with in-home providers to examine their critical role for the continuity of PHC with specific reference to key themes that illustrate their roles, responsibilities and challenges. This leads to a discussion of the underlying research question: what does the inclusion of care in the home mean for the sustainable future of PHC? Finally, we reflect on how our approach contributes to an understanding of who provides PHC and the settings where it occurs, and advance suggestions for ways to extend such research in order to inform PHC policy in Canada and elsewhere.

## Conceptualizing the Home as a Place of PHC

Our approach to understanding the importance of providing care in the home draws on three bodies of literature: (1) geographical perspectives on care and caregiving; (2) feminist critiques of the under-valued nature of caregiving work; and (3) the concepts of continuity and continuum of care and the providers of health human resources. The research in this chapter extends these perspectives to the relatively under-researched context of providers of care in the home and PHC.

Recent work within health geography, and elsewhere in human geography, has highlighted the critical importance of understanding care and caregiving in various settings including the home (Milligan et al. 2007). Within the literature, it is recognized that geographies of caregiving in the home are shaped by health and social care institutions and policies, material structures of the home, health care tasks, schedules, routines and processes, and individual and family characteristics. Key contributions critique and address the gaps in theorizing the spatial relationships and processes involved when the home becomes a place of caregiving (Dyck et al. 2005; Milligan 2001; Wiles 2005). Together, these studies question the dualisms and polarities of paid/unpaid, formal/informal and private/public when homes become places of care, and also critique how restructuring of health care policy and service delivery affect the material, social and emotional conditions of homes as settings for caregiving (Milligan et al. 2007). Such critiques are useful for understanding the geography of PHC as it relates to provisioning in the home.

Underlying the growing concern for understanding the home as a place of care is the recognition that caregiving is a highly gendered activity. For instance, research in Canada (see Michelson and Tepperman 2003), the United Kingdom (see Twigg 2000) and the United States (see Brown 2004) recognizes the taken-for-granted and under-valued work of care providers irrespective of whether the care is paid or unpaid, formal or informal, as predominantly involving women (Jenson and Jacobzone 2000). Feminist scholars have critiqued the under-valued nature of women's care work in the home and how this impacts their lives (Grant et al. 2004). As part of this critique, the literature has shown that the home is usually the first point of contact with the PHC system through the efforts of informal and formal providers as a *de facto* result of downloading in the health care system (Aronson and Neysmith 1997). Providers of care in the home initiate the contact either by calling the physician or calling the home care organization to alert superiors to a potential problem and facilitate the contact by transporting the individual to the health care professional and also by sharing health information that may be essential for proper diagnosis and treatment. The changing context of home and community care means that the distinction between roles and responsibilities in PHC is increasingly blurred (Lewis 1999). The gendered and blurred nature of care in the home is useful for understanding its critical, yet ambiguous role as part of the continuum of PHC.

The changing geographies of care in the home and community mean that in-home providers of care are responsible for more complex, technologically sophisticated and demanding health care work. It is important to broaden the conceptualization of health human resources to include 'all those who contribute to the objectives of the health system, whether or not they have formal health-related training or work in the organized health sector' (WHO 1990, 45). In addition, providers of care in the home have an important role to play in both the continuity and continuum of care.

Continuity of care in PHC is 'mainly viewed as the relationship between a single practitioner and a patient that extends beyond specific episodes of illness or disease' (Haggerty et al. 2003, 1219). The three types of continuity include: (1) informational continuity (i.e., the use of information on past events and personal circumstances to make current care appropriate for each individual); (2) management continuity (i.e., a consistent and coherent approach to the management of a health condition that is responsive to a patient's changing needs); and (3) relational continuity (i.e., an ongoing therapeutic relationship between a patient and one or more providers) (Haggerty et al. 2003). For many individuals, such as children, frail elderly and those with complex disabilities, providers of care in the home, whether they are family members or personal support workers, hold much of the information and knowledge that is essential for establishing continuity of care.

McBryde-Foster and Allen (2005, 630) define continuum of care as 'a series of initiating, continuing and concluding care events that result when the patient seeks providers in one or more environments within the health care system.' In their analysis of the nursing literature, they note that families and caregivers are included in people involved in the continuum of care. In reviewing the descriptions of the environment associated with the continuum of care as part of this analysis, the keyword home was mentioned 188 times and the keyword community was mentioned 160 times (McBryde-Foster and Allen 2005). Continuity and continuum of care, therefore, provide useful points of reference for elucidating the critical roles that providers of care in the home have in PHC.

## Neglect in PHC Policy and Reform

While the importance of PHC provisioning in the home is clear, it remains unrecognized and under-valued within prevailing health care reform policies, and Canada is no exception (Wilson et al. 2004). For instance, according to Health Canada [1] 'Primary health care recognizes the broader determinants of health and includes coordinating, integrating and expanding systems and services to provide more population-based, preventive and health promotion services through the best use of *all care providers, not necessarily those provided only by doctors*' (emphasis added). PHC is important for the direct provision of first-contact services by providers such as family physicians, nurse practitioners, pharmacists and telephone advice lines, and the coordination of continuity of care across the health care system (Health Canada [2]). The PHC system, however, gives little recognition to the importance of unpaid and paid providers of care in the home (Health Council of Canada 2005). This contradicts the World Health organization (WHO) Alma-Ata declaration (1978) which defines PHC as '[E]ssential health care based on practical, scientifically sound and socially acceptable methods and technology made universally accessible to individuals and families in the community through their *full participation*' (emphasis added). Such lack of acknowledgement and support of the active involvement of individuals and families in PHC marginalizes their full participation.

The marginalization of families and individuals, not to mention paid and volunteer providers of care in the home, is also apparent in the concepts of PHC teams and first-contact care. Reform initiatives triumph multidisciplinary PHC teams as '[A] broad range of health care providers who work interdependently in a collaborative partnership to deliver primary health care services with the patient as an integral member of the team' (Health Council of Canada 2005, 8). For instance, the Ontario Ministry of Health and Long-term Care uses the term family health teams and suggests that they '[A]re designed to give doctors support from other complementary professionals, [and] will consist of doctors, nurses, nurse practitioners and other health care professionals.' Key studies, however, critique the narrow conceptualization of PHC team because it ignores health policy and service delivery changes that have expanded the range of individuals and professionals that provide PHC, including all providers of care in the home (see Taket, 2005). Buetow (2004, 509) echoes this call for greater inclusion, arguing that the 'clinician-centric perspective devalues contributions from patients and their informal caregivers and limits understanding of key concepts, their effects and health policies for achieving them.'

Similarly, the first level or point of contact is one of the key concepts in PHC delivery (Health Council of Canada 2005). Many governments, including all provinces and territories in Canada, have instituted health care reforms to try and reduce the number of hospital emergency room visits and admissions, the length of hospital stays and improve waiting lists. These reforms still focus on physicians and their places of employment including offices and after-hours clinics (Buetow 2004). This inevitably neglects the home as a place of PHC and the many activities concerning the prevention, maintenance and promotion of health that occur there. Therefore, if the first point of contact is where the initiation of contact with the health care system occurs, providers of care in the home must be part of conceptualizing and defining PHC. With the widespread policy neglect in mind, we now focus our attention to a specific example of the how providers of care in the home contribute to PHC.

## Chronic Care and the Illness Trajectory

In Canada, and internationally, chronic disease management is a key concern in PHC provision. Sixty percent of deaths worldwide in 2005 were due to chronic diseases including cancer, cardiovascular disease, diabetes, kidney disease and respiratory disease (WHO 2005, 3). Proper chronic disease management can improve the quantity and quality of life and decrease overall health system costs. Providers of care in the home often look after many of the components for effective chronic disease management including coordinating care, minimizing unnecessary visits and admissions, and providing care in the least intensive setting (Health Council of Canada 2005). The home is a key place for the management of chronic health conditions due to an aging population, increasing portability of health care technology, increasing life expectancies for individuals with chronic conditions and policy and personal preference to remain in the home (Wiles 2005).

Chronic care management can be understood in reference to dealing with changes in the illness trajectory. A trajectory refers to 'the total organization of work done over the course of the illness plus the impact on those involved with that work and its organization' (Corbin and Strauss 1988, 34). The work in question encompasses both illness-related work such as managing medications and medical appointments, and biographical work such as coming to terms with changes in health and how to balance illness-related work with everyday life. The required work depends on the type of illness; all chronic illnesses do not require ongoing treatment or necessitate caregiving assistance from others.

The trajectories of many chronic conditions vary between acute, comeback, stable, unstable, downward and dying, and within the lifespan of an individual several or all of these phases are experienced (Corbin and Strauss 1988). The management of some chronic illnesses on a daily basis requires negotiations with many components of the PHC system. Increasing calls for self-management are necessarily employing ill individuals and their caregivers to engage in care practices in the private sphere of the home. Advances in pharmaceuticals and technologies, while allowing for provision of care in the home, require frequent monitoring within the PHC system through coordinated care efforts between medical professionals and formal providers of care in the home (e.g., case managers and home care workers). Contact with the formal PHC system is also needed in order to legitimate claims for assistance in the home such as paid care services.

Changes in the trajectory of an illness often precipitate accessing PHC. For example, if a person's health is unstable and the strategies that have been used to manage their chronic illness are no longer effective, they may require additional medications or treatment to return to a stable phase. Whereas, an individual in a downward or dying trajectory often requires frequent contact with PHC professionals to access necessary equipment or pain medication. The various trajectories involve contact with the PHC system, which may be initiated by the individual themselves or providers of care in the home. Chronic care as seen through the illness trajectory, therefore, provides an excellent example of the increasingly blurred boundary surrounding PHC within which providers of care in the home are situated.

**Provider Perspectives from Ontario**

To illustrate the crucial roles and responsibilities of care provisioning in the home with respect to PHC, we turn to the results of two qualitative studies aimed at understanding the perspectives of home and community care providers in the province of Ontario, Canada (Skinner 2005; Yantzi 2005). Like many other provincial and national jurisdictions, Ontario has undergone more than a decade of health care restructuring including reforms to the funding and delivery of chronic care. At the time of the interviews (2002–2004), the home and community care sector was centred on intermediary agencies known as Community Care Access Centres (CCACs), which had responsibility for administering and managing funds provided by the provincial government. In turn, the CCACs contracted-out service delivery for professional support (e.g., nursing and therapies) and personal and homemaking support to for-profit and not-for-profit businesses based on a competitive bidding process (Baranek et al. 2004). The key implications of the CCAC model for providers of care in the home were the diversion of services and funding from chronic to acute care, which resulted in more people vying for services, more restrictive criteria affecting what services care recipients and their families could receive, and a reduction in the number of hours of support that were available (Cloutier-Fisher and Joseph 2000). Arguably, it is because of such changes and constraints that providers of care in the home now play a crucial role in the continuum and continuity of PHC provision.

The present analysis draws selectively on excerpts from in-depth interviews with directors, case managers, paid staff, volunteers and family members involved in providing care in the home. In separate but parallel attempts to build understanding of home and community care in Ontario, Mark Skinner (2005) conducted interviews with representatives of governments, health institutions and authorities, for-profit and not-for-profit businesses, voluntary organizations, community groups and volunteers providing services for older adults (n=72), and Nicole Yantzi (2005) conducted interviews with family members caring for children with long-term chronic conditions (n=12). While a meta-synthesis of these interviews is beyond the scope of the chapter, we note that both studies share similar approaches and methodologies (i.e., intensive research based on grounded theory). Details of the research are reported in related publications (see Skinner 2005, 2008; Skinner and Rosenberg 2006a, 2006b; Yantzi 2005; Yantzi et al. 2007; Yantzi and Rosenberg 2008).

Taken together, the findings from both suites of interviews present a comprehensive in-home provider perspective of home and community care in Ontario, which complements the user and self-care perspectives established in the literature (see Aronson 2002). For consistency, the excerpts that follow are taken directly from the original transcripts of the research and are identified by the author's initials (MS or NY). The re-analysis of qualitative data is organized around the key themes that substantiate and illustrate the crucial role of providers of care in the home in PHC.

*Partners in PHC*

The shift of health care from hospitals to home and community, and the direct delivery of services in the home required providers to care for individuals with evermore intense, complex and technologically sophisticated needs. This situation is encapsulated in the following description of the dynamics among providers in Ontario, where '[A]s the acute care hospital dumps clients onto the CCAC, that affects us in-home providers, and when we start dumping clients onto community support, that affects them, and then you need more volunteers' (MS). Delegation and downloading also affects family members as one service provider explains '[T]hat is the concept of the CCAC. You teach the task and you delegate as much as possible. If there is a family member or caregiver, then you delegate and minimize [formal] services' (MS). Such excerpts substantiate the main theme emerging from the interviews with the providers of care in the home: that there is a *de facto* presence of providers of care in the home in the continuum of PHC. As is already clear, however, their presence is neither acknowledged nor supported in PHC policy or delivery.

*Initiating and Facilitating First-contact*

In many cases, without providers of care in the home, contacting, transporting, relaying health information and following-up on medical regimes required for effective and efficient PHC would not happen. As reported in both studies, there are many reasons why providers initiate contact with PHC professionals including to gain assistance with referrals for equipment or other types of assessment, forms for eligibility for social and health programs, prescriptions for additional medications and/or advice to manage unstable or downward phases, intravenous therapy for a short or long duration, pain management in palliative phases, accessing specialist care and planning surgeries. In addition, providers of care in the home are often the best suited to initiate contact with medical professionals and the delivery of PHC because of their in-depth knowledge and experience working with the individual. Intimate knowledge of an individual's health leads to an awareness of slight changes that may signal more severe underlying problems. Small changes in behaviour or appearance of an individual such as increasing pain, changes in skin colour, laboured breathing and lack of appetite may be a precursor to a looming problem that must be acted upon.

Determining when changes in the health of an individual necessitate contacting health care professionals is another important role. Providers of care in the home who notice changes in a client's health may call the individual's physician or the local organizer of home care services (i.e., a CCAC case manager). For instance, as noted by a volunteer providing care in Ontario, personal contact allows community members to watch out for each other, which in this case means, 'If my neighbour's curtains aren't up by this hour I will call for help' (MS). Clearly, recognition of changes in someone's health status occurs if the provider is familiar with the individual; however, the importance of the consistency of providers in the home in the PHC continuum requires further exploration. A critique of relational continuity is that caregivers may start feeling as though they know what is best for the care recipient without asking her or him; meanwhile others suggest that the development of longer-term relationships may have therapeutic benefits for both parties.

Providers of care in the home also facilitate continuity by managing health information and appointments. Individuals and family members keep records and relay essential information for treatment and health maintenance. The criticalness of this information is emphasized by one provider's statement that, 'You've got to keep on things so you know if there's anything that needs to be reported to the doctors' (NY). At the same time, the maintenance of health records and preparation of applications for medical equipment can require considerable time and energy, adding to caregiver burden and exhaustion. As one mother states '... my filing cabinet is so thick, you wouldn't believe how thick it is. I have more papers than I ever did when I worked, I mean, and I'm always on the computer, doing letters, I do everything on my own, I do everything' (NY). Individuals also rely on volunteers and family members to drive them to appointments with PHC providers. The ability to assist in transportation can be compromised due to the physical requirement of lifting someone in a wheelchair, especially given the aging volunteer base, the gender sensitivity associated with assisting someone traveling for a specific appointment (e.g., mammogram vs. prostate examinations), or lack of appropriate language skills.

## Delivering Care and Making PHC Decisions

Managing, monitoring and responding to changes in the health of recipients is an important component of the illness-related work discussed above. Dealing with changing illness trajectories is part of providing care in the home and the constant awareness of a potential crisis adds to the difficulty. Family care providers described it as always being on standby, ready to react or make a decision about when to initiate contact with other levels of care. For example, a participant notes

> [S]o I think, it's not only that you're looking after somebody, it's always looking out for the possibility that they're, you have to be monitoring their health. It's like you're a full-time nurse you know? If their health, um, that, when they wake up and scream in the night, are they just bored or wet, or are they getting something serious that you might have to rush them to [name of children's tertiary care centre] or to the pediatrician (NY).

This excerpt also provides evidence of the blurring of roles and responsibilities as this mother referred to herself as a full-time nurse. Conflicting emotions about whether a provider is being too cautious or not vigilant enough also must be balanced, especially when caring for someone who is unable to communicate verbally. One provider shares her concerns as follows,

> There's just so many unknowns when you're looking after medically fragile children. Um, you're really afraid for the future, like, 'Am I doing the right thing', 'Am I being vigilant enough?' We don't, you know, you don't want to rush them to the doctor every, every time they have a little cold, but on the other hand, you know that when they do get sick, they can, things can escalate quickly (NY).

Administering and making decisions about medications at home is a common strategy for the treatment of many chronic conditions. Individuals and family members must decide whether more medications should be given and when it is time to seek medical advice. This ongoing decision making is demonstrated by the following excerpt: '[S]o she has a seizure condition also are we going to decide how many more seizures are we going to allow before we call the doctor and say, "Okay let's up the meds again"' (NY). Some individuals with long-term conditions require intermediate PHC to manage surgeries that are required for the maintenance of their health and functional abilities. One mother captures how this changing trajectory required frequent negotiations between PHC and long-term care when she states '... a year ago he had two rounds of hip surgeries, one on each hip, recovery of six to eight weeks. Um, you know major care. The year before that he had spinal fusion done and that was a three month recovery and major surgery' (NY).

### Being Responsible for All Aspects of Health

As outlined above, providers of care in the home assist with a variety of activities essential to personal support (e.g., bathing, toileting, feeding and dressing), home support (e.g., cooking, cleaning, shopping and banking) and community support (e.g., delivering meals, home maintenance, visiting, transportation and caregiver respite). As a respondent working for a for-profit business asserts, however, these activities are often delegated and downloaded onto volunteers and family members, '[O]nce we are out of the game, the community sector [groups and volunteers] is the end of the line' (MS). In many ways, family members are the ultimate end of the line when it comes to meeting daily health care needs. On a daily basis, family members are responsible for changing and toileting, dressing, brushing teeth, administering medications, entertaining, feeding, bathing and assisting with communication, all of which are necessary for maintaining the social determinants of health of individuals.

Discussions about PHC reform and practice have asserted the importance of social determinants of health. The Alma-Ata declaration states that PHC involves, 'in addition to the health sector, all related sectors of national and community development, in particular agriculture, animal husbandry, food, industry, education, housing, public works, communications and other sectors' (WHO 1978, section VII number 4). Many of the personal, home and community support activities listed are critical for people's social determinants of health according to the Alma-Ata declaration. Therefore, the daily activities that providers of care in the home undertake in part determine care recipients' health.

### Challenges of Providing PHC in the Home

The lack of qualified health professionals in many communities in Ontario, especially family physicians, has resulted in the official designation of medically under-serviced communities. Communities with this designation qualify for subsidies and access to financial and other resources to develop their human health resources. While this is an issue in many urban areas, it is exacerbated in rural and remote areas due to the general lack of health professionals and geographic and economic barriers to accessing PHC (Pong and Pitblado 2005). This critical situation is captured by one provider's explanation that '[S]ome of the smaller communities are always only one doctor or one nurse away from a major crisis' (MS). The rural situation reveals another way that providers of care in the home become a *de facto* part of PHC due to the lack of qualified health professionals and infrastructure. Many people do not have a family physician and are limited in where they can access care. As a provider from rural Ontario explains, 'For many people here, the only services that they can get are through our organizations [CCAC and community support]. *There is no where else they can go. There is very little here; if we don't provide it, they don't get it*' (MS, emphasis added).

Despite the key roles that providers of care in the home have in initiating, preparing for and maintaining first-contact with PHC, these providers feel that their expertise and knowledge is rarely recognized, and this is especially the case when considering their role in the PHC continuum. Several mothers, like this one, note how first-contact services often assume that they are ignorant and not knowledgeable when discussing their children's PHC needs:

> [B]ut when you're dealing with special needs child that can't communicate. Then, you know, they [doctors] just tend to assume that, that you don't know anything, even though you know your child better than anybody and there's only a couple of doctors who understand that ... that the parent knows better than anybody else what his, you know problems are, and can probably diagnose him better than anybody, if it's of a regular nature (NY).

In the end, the lack of acknowledgement is viewed as part of the under-valued role of home and community care in general, which is best left to the words of a provider who comments that '[E]ven given all the changes [in the health care system], we still lack legitimacy' (MS).

## Discussion: A Sustainable Future for PHC in the Home?

The purposeful review of in-depth interviews with providers of care in the home from Ontario illustrates some of the multiple and complex ways in which they are involved in PHC. In sum, the excerpts presented above reveal that in-home providers are directly involved in facilitating first-contact with the PHC system, care delivery and decision-making, meeting recipients' social determinants of health and responding to the challenges facing PHC in the community. While such active and direct roles substantiate the literature recognizing the home as an important place of care (see Dyck et al. 2005) and the full range of in-home providers as part of the continuity of care (see Buetow 2004), the evident vulnerability of in-home provisioning raises four questions about what its neglect within prevailing policies and programs means for the sustainability of PHC.

Firstly, in the Canadian context, successful implementation of PHC reform is supposed to strengthen the continuity and continuum of care and achieve a more integrated health care delivery system, change the focus of health care from providers to people and place a stronger emphasis on health promotion and disease prevention (Health Council of Canada 2005). The in-home provider perspective gives decisive evidence that their activities contribute to all of these objectives;

however, since this does not translate into direct support for in-home provisioning, how can such policies be fully realized?

Secondly, the provider perspective gives evidence of the complex ways in which they contribute to continuity and continuum of care. The clearest example is related to the health information that providers of care in the home collect, manage and share with PHC professionals. Examples of the importance of this sharing of information include a mother using it to ensure that her PHC physician has the most recent reports and evaluations for her child, and a personal support worker calling her supervisor concerning changes in the behaviour of her client who in turn contacts the family physician who in turn contacts the emergency room. In fact the goal of making 'current care appropriate for each individual' can only be achieved with the involvement of providers of care in the home (Haggerty et al. 1220). Clearly, the examples of provider activities in Ontario are integral to the continuity and continuum of care, so why do PHC policies and programs remain focused on health professionals working in institutional settings?

Thirdly, given their active and direct roles, it is clear that the success of PHC reform rests, in part, with providers of care in the home. The provider perspective, however, highlights the pervasive lack of legitimacy in terms of the conventional definitions of PHC, PHC teams and first-contact. The exclusion of in-home provisioning within PHC policies and programs means that their crucial role is neither valued nor supported in terms of funding, organization or service delivery. In order for PHC reform to be successful, and for multidisciplinary teams to work, why is there not greater inclusiveness in defining, conceptualizing and theorizing the roles and relationships that comprise the PHC continuum?

Fourthly, the lack of qualified health professionals in many communities, and continued downloading of PHC responsibilities onto volunteers, families and individuals places a tremendous burden on in-home providers, who often do not have the capacity to fulfill their expanding roles (Skinner 2008). A balance must be reached between including providers of care in the home on the one hand, and addressing their vulnerability in the face of taking on ever more responsibility on the other. Indeed, the provider perspective cautions that the physical, social and emotional costs of caregiving must be weighed when adding to the responsibilities of individuals, families and volunteers (Yantzi et al. 2007). Ultimately, if such vulnerabilities are not taken into account, how sustainable is PHC in the home?

## Conclusions

In this chapter, a purposeful review of the results from two qualitative studies has been used to elucidate the important, yet unrecognized roles and responsibilities that providers of care in the home have in PHC. Our purpose was to show, from a provider perspective, that care provisioning in the home must be included in conceptualizations of PHC. Our approach focused on the vulnerability of providers of chronic care in the home in terms of caregiver burnout and lack of legitimacy. We contend that the inclusion of providers of care in the home within PHC policy, research and practice is incomplete without recognition of their ambiguous position within the formal health care system and the attendant paradox of society's growing reliance on these under-resourced and, in the longer-term, potentially unsustainable partners in PHC.

We view the contribution of the research presented in this chapter as three-fold. It addresses recent calls in the literature to broaden the definition of human health resources and the health care continuum to include all providers of PHC regardless of place or person (Buetow 2005; Pong 1999) and, in doing so, points out that calls for greater inclusion are irrelevant without actual policies, programs and funding to help providers fulfill their critical roles. It highlights how current policy and service trends are forcing providers of care in the home to assume PHC responsibilities that are not recognized in reform policies or programs, which places these providers and their care recipients in increasingly precarious positions. Ultimately, it argues that understanding care provisioning in the home in terms of vulnerability and sustainability is a prerequisite for the development of informed PHC policy.

Absent from our analysis are the perspectives of PHC users and recipients, which in an era of health promotion and emphasis on self-care, is an important dimension to consider. In addition, in this version of our work, we have not addressed the gender aspect of care in the home; although this is presented elsewhere (see Yantzi and Rosenberg 2008). Despite these limitations, we view the research as being of considerable analytical interest for policy, research and practice in other provincial and national jurisdictions undergoing similar processes of PHC reform. In closing, we observe three areas of further study to critically evaluate the increasing reliance on providers of care in the home within PHC. Firstly, it is important to better understand how the activities of various providers are essential for meeting the objectives of reforms. Further analyses that differentiate among the providers and the types of care in the home are required (i.e., beyond chronic). Secondly, restrictions in the types of tasks that paid providers are permitted to do result in a greater reliance on family members and volunteers to meet individuals' complex care needs. Examining how these activities promote and meet the broader social determinants of health would substantiate the call for the inclusion of all providers of PHC. Finally, moving beyond inclusiveness and towards supporting providers directly will require further analyses of their roles in ensuring continuity of care across the health care system. It is here that the chapter along with the proposed areas of study can inform policy that comprises and sustains all providers of PHC.

## Acknowledgements

The chapter draws on research funded by the Canadian Institutes of Health Research, Canadian Health Services Research Foundation, Social Sciences and

Humanities Research Council of Canada, and The Hospital for Sick Children Foundation. We are grateful for the invaluable contributions of the anonymous interview participants from Ontario. The views expressed are, however, exclusively our own.

## References

Aronson, J. (2002), 'Frail and Disabled Users of Home Care: Confident Consumers or Disentitled Citizens', *Canadian Journal on Aging* 21:1, 11–25.

Aronson, J. and Neysmith, S.M. (1997), 'The Retreat of the State and Long-term Care Provision: Implications for Frail Elderly People, Unpaid Family Carers and Paid Long-term Care Workers', *Studies in Political Economy* 53, 37–66.

Baranek, P.M., Deber, R.B. and Williams, A.P. (2004), *Almost Home: Reforming Home and Community Care in Ontario* (Toronto: University of Toronto Press).

Brown, M. (2004), 'Between Neoliberalism and Cultural Conservatism: Spatial Divisions and Multiplications of Hospice Labor in the United States', *Gender, Place and Culture* 11:1, 67–82.

Buetow, S.A. (2004), 'Towards a New Understanding of Provider Continuity', *Annals of Family Medicine* 2:5, 509–11.

Buetow, S.A. (2005), 'To Care is to Coprovide', *Annals of Family Medicine* 3:6, 553–5.

Cloutier-Fisher, D. and Joseph, A.E. (2000), 'Long-term Care Restructuring in Rural Ontario: Retrieving Community Service User and Provider Narratives', *Social Science & Medicine* 50, 1037–45.

Cooper, C., Coward, M. and Lessing, M.P.A. (2003), 'An Outpatient Home Parenteral Antibiotic Service', *Primary Health Care* 13:8, 21–2.

Corbin, J.M. and Strauss, A. (1988), *Unending Work and Care: Managing Chronic Illness at Home* (San Francisco: Jossey-Bass Publishers).

Dyck, I., Kontos, P., Angus, J. and McKeever, P. (2005), 'The Home as a Site of Long-term Care: Meanings and Management of Bodies and Spaces', *Health & Place* 11, 173–85.

Grant, K.R, Amartunga, C. and Armstrong, P. et al. (eds) (2004), *Caring For/ Caring About: Women, Homecare, and Unpaid Caregiving* (Toronto: Garamond Press).

Haggerty, J.L., Reid, R.J. and Freeman, G.K. et al. (2003), 'Continuity of Care: A Multidisciplinary Review', *British Medical Journal* 327, 1219–19.

Health Council of Canada. (2005), *Primary Health Care: Background Paper* (Toronto: Health Council of Canada).

Jenson, J. and Jacobzone, S. (2000), 'Care Allowances for the Frail Elderly and their Impact on Women Care-givers', *OECD Labor Market and Social Policy Occasional Papers, No. 41* (Paris: OECD Publishing).

Lewis, J. (1999), 'The Concepts of Community Care and Primary Care in the UK: The 1960s to the 1990s', *Health and Social Care in the Community* 7:5, 330–41.

McBryde-Foster, M. and Allen, T. (2005), 'The Continuum of Care: A Concept Development Study', *Journal of Advanced Nursing* 50:6, 624–32.

Michelson, W. and Tepperman, L. (2003), 'Focus on Home: What Time Use Data Can Tell About Caregiving to Adults', *Journal of Social Issues* 59:3, 591–610.

Milligan, C. (2001), *Geographies of Care: Space, Place and the Voluntary Sector* (Aldershot: Ashgate).

Milligan, C., Atkinson, S., Skinner, M.W. and Wiles, J. (2007), 'Geographies of Care: A Commentary', *New Zealand Geographer* 63, 135–45.

Pong, R.W. (1999), 'Preparing for Population Aging: In Search of a New Health Manpower Paradigm' in Proceedings of the Asia-Pacific Regional Conference for the International Year of Older Persons, Vol. 1, Hong Kong (Hong Kong Council of Social Service), 11–23.

Pong, R.W. and Pitblado, J.R. (2006), *Geographic Distribution of Physicians in Canada: Beyond How Many and Where* (Ottawa: Canadian Institute of Health Information).

Skinner, M.W. (2005), *Voluntarism and Long-term Care in the Countryside: Exploring the Implications of Health Care Restructuring for Voluntary Sector Providers in Rural Ontario (1995–2003)* (Unpublished PhD dissertation, Department of Geography, Queen's University, Kingston).

Skinner, M.W. (2008), 'Voluntarism and Long-term Care in the Countryside: The Paradox of a Threadbare Sector', *The Canadian Geographer* 52:2, 188–203.

Skinner, M.W. and Joseph, A.E. (2007), 'The Evolving Role of Voluntarism in Ageing Rural Communities', *New Zealand Geographer* 63, 119–29.

Skinner, M.W. and Rosenberg, M.W. (2006a), 'Informal and Voluntary Care in Canada: Caught in the Act?', in C. Milligan and D. Conradson (eds), *Landscapes of Voluntarism: New Spaces of Health, Welfare and Governance* (Bristol: Policy Press), 91–114.

Skinner, M.W. and Rosenberg M.W. (2006b), 'Managing Competition in the Countryside: Non-profit and For-profit Perceptions of Long-term Care in Rural Ontario', *Social Science & Medicine* 63, 2864–76.

Taket, A. (2005), 'Primary Health Care in the Community', in D. Sines, F. Appleby and M. Frost (eds), *Community Health Care Nursing*, 3rd edition (Oxford: Blackwell), 26–40.

Twigg, J. (2000), *Bathing-the-Body and Community Care* (New York: Routledge).

Wiles, J. (2005), 'Home as a New Site of Care Provision and Consumption', in G.J. Andrews and D.R. Phillips (eds), *Ageing and Place: Perspectives, Policy, Practice* (New York: Routledge), 79–97.

Wilson, R., Shortt, S.E.D. and Dorland, S. (eds) (2004), *Implementing Primary Care Reform: Barriers and Facilitators* (Kingston and Montreal: McGill-Queen's University Press).

World Health Organization (1978), *Declaration of Alma-Ata International Conference on Primary Health Care* (World Health Organization).

World Health Organization (1990), *Coordinating Health and Human Resources Development* (Technical Report Series 801) (World Health Organization).

World Health Organization (2005), *Preventing Chronic Diseases: A Vital Investment* (World Health Organization).

Yantzi, N.M. (2005), *Balancing and Negotiating the Home as a Place of Caring: The Experiences of Families Caring for Children with Long-term Care Needs* (Unpublished PhD dissertation, Department of Geography, Queen's University, Kingston).

Yantzi, N.M. and Rosenberg, M.W. (2008), The Contested Meanings of Home for Women Caring for Children with Long-term Care Needs in Ontario, Canada', *Gender, Place and Culture* 15:3, 301-315.

Yantzi, N.M., Rosenberg, M.W. and McKeever, P. (2007), 'Getting Out of the House: The Challenges Mothers Face when Children have Long-term Care Needs', *Health and Social Care in the Community* 15:1, 45–55.

**Internet-based References**

Health Canada [1] 'Sharing the Learning: Health Transition Fund: Synthesis Series: Primary Health Care' (updated February 28, 2006) http://www.hc-sc.gc.ca/hcs-sss/pubs/acces/2002-htf-fass-prim/sum-som_e.html, accessed June 27, 2007.

Health Canada [2] 'About Primary Health Care' (updated June 21, 2006) http://www.hc-sc.gc.ca/hcs-sss/prim/about-apropos/index_e.html, accessed May 28, 2007.

Ontario Ministry of Health and Long-term Care 'Family Health Teams', http://www.health.gov.on.ca/transformation/fht/fht_understanding.html, accessed July 20, 2007.

Chapter 14

# On the Street: Primary Health Care for Difficult to Reach Populations

David Conradson and Graham Moon

Since the formation of the National Health Service (NHS) in 1948, successive British governments have sought to ensure equity of access to primary health care (PHC). The traditional gateways to the PHC system, the general practitioner (GP) and the Accident and Emergency (A&E) department, have accordingly been subject to regular reviews and periodic reform (Chapman et al. 2004; Majeed et al. 1994). Supplementary sites for provision, such as community clinics and pharmacies, have been developed and in recent years telephone and internet-based information services have also been introduced. Although PHC provision is still largely associated with the public sector, with the NHS, it is no longer exclusively so. Both voluntary and private sector agencies are now involved in delivering PHC. Looking beyond conventional medical services, we can also identify a broader primary care commitment through the provisions of the welfare state, within social care organizations, and in the work of family and individual carers.

Our interest here is in PHC service organizations and their engagement with disadvantaged populations. Significant inequities in access to PHC exist within contemporary Britain (Townsend and Davidson 1982; Townsend, Davidson and Whitehead 1992; Department of Health 2005). Analysis of NHS data suggests that low income and non-white groups have higher than anticipated GP consultation rates (relative to need), but that there are also important variations within these communities (Morris, Sutton and Gravelle 2005; Smaje and Le Grand 1997). In particular, disaggregation reveals a number of so-called 'difficult to reach groups' whose use of PHC services is in fact below expectation. Among the most well known cases are vulnerable migrants, people who are homeless, intravenous drug users, and street sex workers (Morris et al. 2005).

We examine two forms of PHC service in relation to these difficult to reach groups. The first is the NHS Walk-in Centre (WiC), a form of nurse-led facility introduced across Britain in 2000. The widely publicized purpose of WiCs has been to address 'issues of access and unmet need in primary care, and to meet public expectations about the delivery of modern primary care services' (Pope et al. 2005, 40). By offering longer opening hours and consultations without prior appointment, the intention has been to make PHC accessible to a broader cross-section of the United Kingdom (UK) population. In late 2007 there were around 90 WiCs in the UK and recent health expenditure commitments indicate a possible increase to around 150 by 2010 (Walker 2007). Second, we consider non-statutory initiatives that seek to address the health care needs of people who are homeless, with a particular focus on voluntary sector organizations. Although a diverse social group, homeless people are well recognized as being difficult to reach through conventional PHC initiatives (Daiski 2007; Schanzer et al. 2007; Wen, Hudak and Hwang 2007).

Our discussion draws on examples from the cities of Portsmouth and Southampton in southern England. Both the WiCs and non-statutory organizations in these cities seek to overcome the barriers some individuals experience with respect to accessing traditional health care sites, such as the GP clinic. As we demonstrate, however, the patient groups reached by each type of service differ considerably. There also remain individuals whose needs might be more effectively met by proactive outreach services, where health professionals engage with people in their homes or, in the case of homeless individuals, perhaps in a shelter or literally on the street. Geography matters in both our examples, and not only in terms of physical proximity to consumers. We explore the location of services with respect to the populations in need, whilst also noting that the very existence of such provision, as well as its nature, is contingent upon place specific intersections between human initiative, financial resources and policy imperatives.

The chapter has three main sections. First, we review conceptualizations of access to PHC, highlighting the difficulties of difficult to reach groups in this regard. Second, we consider the contemporary landscape of PHC provision in the UK, noting both longstanding and more recent developments. Drawing on examples from Portsmouth and Southampton, the third section then compares NHS WiCs with voluntary sector PHC services, discussing their strengths and limitations with regard to the health care needs of disadvantaged populations. A short conclusion follows.

## Access to Primary Health Care: The Experience of Difficult to Reach Groups

Equity of access is an important objective in much PHC policy (Gulliford and Morgan 2003). It is also a complex phenomenon to define and measure. We know that utilization varies by social factors such as gender, age, ethnicity and employment category (Chapman et al. 2004; Dixon et al. 2003), for instance, but whether such variation can be deemed problematic depends upon its relation to need.

In seeking to understand social and geographic variations in the utilization of PHC services, both the availability of a service (its supply) and those factors which determine its uptake are important (Dixon et al. 2003). Variations in service availability might be measured in terms of the number of doctors or hospitals per unit of population, or the average travel time from a given neighborhood to an A&E facility (Joseph and Phillips 1984; Gatrell 2002). Whether a local health care facility has the *capacity* to take on new patients is also important: physical existence of a service cannot be equated with its availability. This is a particularly pertinent issue with respect to NHS dentistry in the UK, as practices have only a limited number of publicly funded units of dental provision. Beyond this, all services are charged at private rates. So whilst a community may possess a NHS dental practice, accessing publicly subsidized dental care from it is typically very difficult, as demand greatly exceeds supply in most places.

With respect to uptake, the human response to available PHC services is a major determinant of utilization (Adamson et al. 2003). Voluntary non-use is a significant matter, and not only for busy professionals who forgo medical appointments because of work or travel commitments. People from disadvantaged social groups, such as vulnerable migrants and the homeless, may also choose not to use local PHC facilities. Past experiences of stereotyping or perceived stigmatization have been implicated in deferred visits or even systematic avoidance of PHC services (Wen et al. 2007). A more general discomfort with the formality, power inequalities or personal disclosure associated with clinical consultations may also discourage utilization (Daiski 2007).

In the UK, the intersection between these supply and demand factors contributes to 'inequalities in utilization with respect to income, ethnicity, employment status and education' (Morris et al. 2005, 1251). More specifically, Majeed et al. (1994, 226) note that '[y]oung children, elderly people, residents of socially deprived areas, and members of ethnic minorities have a greater need for care and have higher than average consultation rates.' Despite the generally pro-poor composition of PHC visits, however, there remain marginalized groups that are under-represented (Blumenthal, Mort and Edwards 1995). As noted earlier, these include vulnerable migrants, people who are homeless, intravenous drug users and sex workers. The reasons for under-utilization of PHC services differs in each case, but homeless people are often negatively affected by the absence of a stable residential address, and sex workers may be concerned about moral judgements regarding their lifestyles.

The literature offers several windows onto the health care needs and experiences of these individuals. Amongst vulnerable migrants, the economic precariousness and improvised nature of their everyday lives often leads to 'difficulty fitting in with the way mainstream health services are organized' (McColl, Pickworth and Raymond 2006, 116). There may be minimal overlap between non-work time and service's opening hours, or getting to a clinic may involve onerous public transport journeys. There is then the challenge of receiving advice and information in what is often not a migrant's first language (Feldman 2006). In addition, restrictions have been placed upon migrants' access to some elements of hospital-based PHC in Britain since April 2004. Such constraints run in parallel with welfare initiatives that also seek to limit the impact of immigration on public services and infrastructure.

Amongst homeless people, morbidity is significantly worse than in the general population and, in the UK, death rates can be up to 25 times higher than for housed individuals (Shaw, Dorling and Brimblecombe 1999). Such figures speak of a broader material and psychosocial vulnerability, with many homeless people experiencing multi-faceted difficulties with 'physical illnesses, mental health, addictions and stress' (Daiski 2007, 273). The prevalence of violence in shelters and on the streets, and the anxiety and emotional distress associated with having no fixed abode may also be problematic. Compounding these difficulties is the observed reticence among homeless people to engage with the formal health care system (Quilgars and Pleace 2003). Some individuals have found their previous interactions sufficiently dehumanizing that they largely avoid seeking further PHC assistance (Wen et al. 2007).

In summary, there is considerable evidence regarding the health needs of disadvantaged groups. Moreover, it is clear that such needs are often compounded by difficulties with mainstream care provision: problems of access, comprehension, compliance and affordability. These difficulties reflect both structural and behavioural factors and implicate both service users and providers. On the margins of society, among excluded populations, the difficulties can often be further intensified. The expectations and requirements associated with use of the mainstream PHC system are often incompatible with fractured, transient and troubled lives.[1]

## The UK Landscape of Primary Health Care Services

Having considered the health care experiences of difficult to reach groups, we now examine policy and practice concerning the enhancement of access to PHC. We begin by reflecting on the policy recognition that PHC is more than a service delivered by registered health practitioners in traditional settings, most notably the clinic or surgery (Moon and North 2000). We then provide an outline typology of PHC services in the UK, considering both traditional and more recent initiatives that have sought to address the needs of disadvantaged groups.

The British government seeks to address the PHC needs of disadvantaged populations through a linked set of policy imperatives. In general terms these policies are rooted in some formulation of welfarism. Even during periods of strong neo-liberalism, states generally show some concern for the welfare, including the health, of the disadvantaged. Many put safety-net schemes in place to ensure access to a basic package of care, such as the Medicaid programme in the United States. Voluntary organizations often then fill services gaps. Much the same is true in countries, such as Canada, with a mix of public and private funding of health

---

1  We note in passing that 'trouble' with mainstream PHC is not confined to the materially poor and socially excluded. There are well-off, non-disadvantaged populations who also experience difficulties with the more rigid aspects of mainstream PHC provision in the UK. For these populations, the exigencies of a successful (post)modern existence prove a barrier to accessing PHC. Fragmented, busy and speedy lives may not always mesh well with relatively fixed, routinized health care provision.

care. A well-developed welfarist position is most evident in tax-based national health services where, as for example in Britain, there is often an overt policy commitment to provide health care for all according to need. In recent years, the election of successive New Labour governments in the UK has ensured that this historic policy commitment has been accompanied by an equally relevant concern to combat social exclusion and address the needs of marginalized population groups (Hills and Stewart 2005; Pantazis, Gordon and Levitas 2006). Legislation has made it clear that Primary Care Trusts (PCTs) have a responsibility for all citizens within their catchment areas. At the same time, governmental commitment to flexibility and modernization has challenged the 'one size fits all' approach to health care provision, with a key aim being to enhance supply and accessibility.

Converting this policy agenda into action has not proved easy. In the UK, NHS expenditure has risen markedly since 2000, but much of the increase has been absorbed by efforts to improve the quality of existing PHC arrangements rather than extending services to disenfranchised groups. Where extension has happened, it has had four key characteristics. First, where it has remained NHS-led, it has entailed restructuring traditional divisions of labour. Both nurses and pharmacists have taken on roles formerly reserved for medical doctors, for example, producing a more flexible health care workforce, often better placed to communicate with users in a wider variety of geographical settings. The spatial equation of PHC with the GP surgery has thus been fractured. Second, although there is a long history of self-help and volunteer led initiatives operating alongside the NHS, the neo-liberal preference for reduced public sector involvement has fostered a rise in non-statutory provision of health care for disadvantaged groups. There is now a range of businesses, charitable schemes, and partnerships between voluntary and public organizations that provide such health services. Third, there have been concerted moves to ensure better temporal availability of PHC. Opening hours have been extended and the range of out-of-hours services has increased as the variety of service providers has grown. Finally, it has been recognized that the UK requirement to register with a single general medical practice, with that practice functioning as a gateway to both primary and secondary health care, is not always realistic. Historically some 5 percent of the population have not registered in this way, and socially disadvantaged groups have often featured prominently in this regard. While safety nets have always existed and non-registered populations have developed strategies for accessing health care, recent initiatives have lessened the need to be registered with a single practice.

The outcome of these policy developments is a complex landscape of PHC provision, encompassing both traditional and more recent initiatives (see Table 14.1). General medical practice remains a key element, not least as it continues to provide the majority of PHC services, but it is no longer a uniform operation. There are now multiple arrangements for the provision of care outside standard working hours, involving cooperatives, rota systems, and contracting out to private or NHS practices (HCHC 2004). This diversity reflects the new general practice contract introduced in 2004, under which many surgeries now no longer provide out-of-hours care, having instead outsourced it to private sector operators. This decision has been aided by a shift from a funding system in which general medical practices essentially functioned as small businesses to one in which an increasing number of practices are staffed by doctors drawing a salary from the local PCT. This shift has enabled PCTs to direct primary medical care more effectively. They are better placed to ensure the provision of out-of-hours care, as well as to facilitate the growth of specialist clinics and medical practices for disadvantaged and socially excluded populations.

**Table 14.1  Primary health care services relevant to disadvantaged populations in the UK**

| Traditional NHS Providers | Recent NHS Developments | Non-Statutory Initiatives |
|---|---|---|
| • General Practice Surgeries<br>• Community Health Clinics<br>• Accident and Emergency Departments<br>• Community Pharmacies | • NHS Direct (2000-)<br>• NHS Walk-in Centres (2000- )<br>• Specialist clinics under the Personal Medical Services initiative (1998-) | • Voluntary and community sector projects<br>• Private sector telephone and advice centres |

Beyond general medical practice but still within the NHS, it is possible to identify a number of long-standing services that also provide PHC to disadvantaged populations. First, PCTs and their predecessor equivalents have a long history of funding community clinics. Often directed at groups with specialist needs or people not registered with general medical practices, these services represented the historic safety net for disadvantaged individuals seeking PHC. Many have now been superseded by personal medical services or PCT medical service provision, or indeed by nurse-led care. More recently such safety-net provision has, in any case, often been delivered by a second long-standing service: the A&E department of the local hospital. In the 1980s, a review of PHC in London noted the particular tendency for A&E departments in inner city areas to function as surrogate PHC provision (LHPC 1981). For excluded and marginalized populations, this phenomenon is marked and perhaps understandable not only in terms of the difficulties entailed in accessing general medical practice but also, consequentially, because non-contact with general medical care may cause health problems to deteriorate to a state necessitating emergency treatment. Despite other developments, the unintended use of A&E departments as PHC providers remains a problem. A third long-standing NHS service, now being extended to provide PHC, is the community pharmacy (DoH 2000). With an established retail presence in many high streets, these are typically a more spatially extensive form of provision than general practice surgeries. Many common chronic ailments can be treated with over-the-counter medications and socially excluded groups traditionally make substantial use of pharmacies, particularly those with extended opening hours. Although community pharmacies thus seem an ideal site for the delivery of PHC to marginalized populations, they remain a professionalized, white coat service rather than a more genuinely street-level intervention.

Having described long-standing NHS services, we now consider recent NHS efforts to enhance access to PHC. One such service, NHS Direct, represents a technological fix. Available as a telephone and internet-based advice service, NHS Direct is largely nurse-delivered and seeks to exploit developments in telecommunications to ensure round-the-clock provision. It is clearly promoted as primarily an advice service, with users instructed to contact their GPs if in need of a diagnosis. Nevertheless some of the information provided enables self-diagnosis and treatment. A number of similar private sector services exist, although some offer significantly more in the way of diagnosis. In terms of our interest in difficult to reach populations, NHS Direct and its private equivalents are of limited relevance. They depend on access to communication technologies and, though this is now widespread, penetration is least amongst those in greatest need and consequently use of NHS Direct declines among the most marginalized populations (Burt, Hooper and Jessop 2003). Its utility is also compromised by its status as an advice rather than treatment service.

Such shortcomings do not apply to a second service innovation: the NHS WiC. Operating on a drop-in basis, with extended opening hours, these nurse-led centres provide both advice and treatment. The aim is to supplement general practice and remove PHC consultations from A&E departments. Over 90 WiCs have been opened since 2000 and, at least in theory, they should be accessible to otherwise excluded populations. Although some WiCs have been deliberately located in deprived areas, evidence suggests that their users are often drawn from more affluent groups (Salisbury et al. 2002; Salisbury 2007). The vast majority of users are also already registered with a GP, suggesting that the WiCs may operate as a complement to mainstream PHC rather than successfully targeting marginalized populations that are not registered with a doctor.

Moving beyond established and recent NHS services brings us to the realm of non-statutory initiatives. Here we find special projects, charitable schemes and other bespoke provision for marginalized and excluded groups. Such services typically originate in the voluntary and community sector, but may later become hybridized as a result of cross-sectoral funding partnerships or processes of organizational evolution. There is some relevant private sector provision in this category, such as advice services, but on the whole private agencies are oriented towards affluent clients.[2] PCTs may provide some coordination and funding for this patchwork of provision, particularly through the personal medical services

---

2   The private sector equivalents of NHS WiCs, for instance, are less relevant to a consideration of disadvantaged populations. Their intended patient group typically has

initiative.[3] In any case, certain charities have developed considerable expertise in delivering PHC. For example, St. Mungo's, the London homelessness agency has developed early intervention services at its hostels.[4] In their review of health care for homeless people, Quilgars and Pleace (2003) confirm both the range of services and the strong role played by such localized schemes. They also emphasize the barriers to care that result from negative societal attitudes, the requirement for a permanent address or registration with a GP, and the problematic trade-off many homeless people are forced to make between immediate survival needs and longer term health maintenance (see also Daiski 2007; McColl et al. 2006; Schanzer et al. 2007). Voluntary sector agencies are arguably well positioned to address some of these problems, although their presence is generally contingent upon local initiative and stable funding. In addition, they may not be well integrated with other providers.

Despite these limitations, one might argue that PHC services for the most marginal and excluded members of British society are potentially delivered most effectively by voluntary sector schemes and projects. Traditional forms of NHS provision certainly all have limitations of one form or another. Yet recent NHS developments, such as WiCs and the growth of personal medical services, clearly offer possibilities for improved public sector delivery of health care to disadvantaged populations. We now turn to two case studies that examine these issues.

## Primary Health Care Services in Portsmouth and Southampton

Portsmouth and Southampton are located on the south coast of England, some 100km south west of London. They are both medium sized university cities – the 2001 UK census recorded a population of 217,500 for Southampton and 186,700 for Portsmouth – with shared but distinct maritime heritages, now partially supplanted by service industries. Portsmouth is traditionally associated with the British Navy and retains a substantial naval dockyard; Southampton's historical associations are with commercial shipping and both cruise and container vessels are now major port users. It is tempting but not entirely accurate to link the relatively high levels of homelessness in both cities to the transient nature of maritime and dock employment. Certainly, however, and particularly in Portsmouth, the

---

private medical insurance or is sufficiently well-off to pay for consultations directly. Such services are consequently not listed in Table 14.1.

3    Introduced in 1998, the Personal Medical Services initiative is a 'locally-agreed alternative to General Medical Service (GMS) for providers of general practice' (https://bmahouse.org.uk/ap.nsf/Content/pmsagreements0904). It offers both NHS and non-statutory providers increased freedom in how they respond to local primary care needs, with the aim of enabling innovative and more locally attuned provision.

4    See http://www.mungos.org/facts/health.shtml (accessed 15 September 2007).

manual blue-collar nature of much traditional employment has led to relatively high levels of deprivation in otherwise affluent southern England. In overall terms, Southampton is the 96th most deprived city in England, and Portsmouth ranks 88th (ODPM 2004). Each city has areas amongst the 5 percent most deprived in the country, with Portsmouth possessing the four most deprived neighbourhoods in the county of Hampshire within which both cities are located.

Portsmouth and Southampton each have their own PCT with responsibility for health care delivery. Interviews, site visits and an assessment of documentary material were used to compare and contrast two forms of PHC provision for marginalized populations in these cities. The first were the NHS WiCs and the second were voluntary sector initiatives for homeless people. Figure 14.1 shows the location of the case study facilities, with provision overlaid on a map of local variations in the 2004 index of multiple deprivation (IMD2004), an established measure of community disadvantage. Darker shades, indicating higher IMD2004 scores, denote higher levels of deprivation. As an initial comment, it is clear that provision in Portsmouth is less extensive and generally located peripherally with respect to the areas of greatest disadvantage.

*Walk-in Centres*

The Portsmouth NHS WiC opened in December 2005 in a new building erected on the car park of the city-based St Mary's Hospital (see Figure 14.2). It shares its premises with an NHS treatment centre designed to enhance local capacity for minor surgery. The WiC's catchment is envisaged as not only the city of Portsmouth but also the surrounding area; the WiC itself is relatively peripheral to the city's main transport routes and deprived neighhourhoods. Services offered include diagnostics, a minor injuries unit, health information, and advice and treatment for a range of minor illnesses. This provision is available from 8:00am to 9:30pm, seven days per week, with no appointments necessary and a target of providing an initial assessment within 15 minutes of arrival. The Centre is nurse-led and was developed as a partnership between Portsmouth PCT and Mercury Health, an independent clinical services business. Mercury's literature describes itself as an organization that designs, builds, manages and delivers health care mainly for the NHS: 'free at the point of delivery, in traditional NHS locations and in places where NHS facilities are not available' (Mercury Health 2007a).

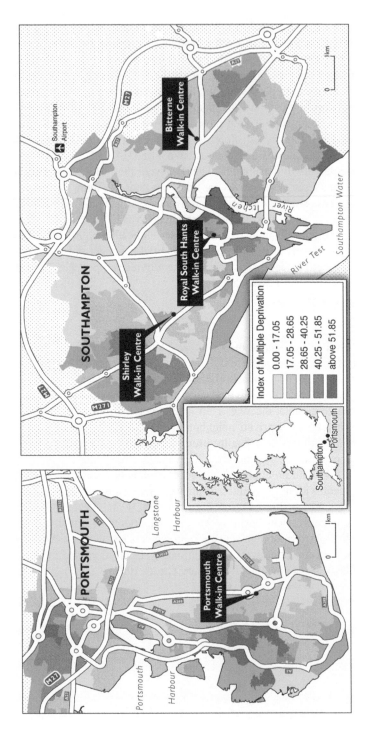

Figure 14.1  Walk-in centres in Portsmouth and Southampton, 2008

**Figure 14.2  St Mary's treatment walk-in centre, Portsmouth**
*Source*: Laura Moon.

Guidance for users makes it clear that the Centre is not for life threatening conditions. These are seen as being more appropriately dealt with by the A&E department at the main local hospital, situated some three kilometres north of the WiC. Patients are expected to attend with 'new or unanticipated general health problems' (Mercury 2007a). In this sense, the Centre seeks to provide additional local choice for people with PHC needs. It forms part of a network of NHS services in which '[t]he nursing staff will thoroughly assess you and if they are not able to treat you will refer you on to your GP, A&E or another appropriate health service' (Mercury 2007a).

Southampton has three WiCs. One (Shirley) is in a refurbished shop premises whilst a second (Bitterne) occupies part of a community health centre (see Figures 14.3 and 14.4). Both see over 100 patients every day. The third and newest is, like the Portsmouth WiC, located on a hospital site where there is otherwise no A&E provision: the Fanshawe Wing of the inner city Royal South Hampshire Hospital. In terms of location, the three centres are relatively well-positioned to serve more deprived populations and to allow people to take advantage of transport links. Each adheres to the standard model of nurse-led care, with a focus on health promotion and treatment of minor injuries and illnesses, and onward referral as appropriate. All have extended opening hours.

**Figure 14.3  Shirley walk-in centre, Southampton**
*Source*: Author's photo.

As in Portsmouth, WiCs are an integral part of the network of NHS care in Southampton. The phased development of the three centres was intended to enhance PHC accessibility, whilst at the same time reducing what is seen as misuse of A&E services in the city. As the service manager for unscheduled care explained at the opening of the third centre:

> Our campaign will hopefully raise awareness of what Walk-in Centres can offer and help people understand how important it is to use NHS resources appropriately, and that being injured or unwell doesn't mean going to the Emergency Department. Emergency Department staff are there to deal with more serious or life-threatening injuries and illnesses (SPCT, 2007).

The centres were promoted as giving 'people more convenient access to NHS services, by providing professional nurse-led care to match modern lifestyles that don't always make it possible to visit a GP during normal working hours. They also provide services to people who are not near or do not have their own GP' (SPCT 2007).

**Figure 14.4  Bitterne health and walk-in centre, Southampton**
*Source*: Author's photo.

Comparing PHC provision in the form of WiCs in each city reveals a number of interesting distinctions. First, as suggested, provision in Southampton is more appropriately located with respect to enhancing access for disadvantaged groups, many of whom rely on walking and public transport for local travel. The central facility in particular is well-placed with respect to more marginalized populations. Second, although both cities make much of the integration of WiCs with other forms of PHC provision and with A&E care, it is only in Southampton that there is explicit recognition of possible users who are not registered with a GP. In the Portsmouth case, much of the website material seems to assume registration. Yet both cities have significant populations that might be expected to have low levels of registration – homeless people, rough sleepers,[5] asylum seekers and recent migrants from Eastern Europe. Third, we have an intriguing contrast between direct provision by a PCT in Southampton and NHS-funded provision via a private sector contract in Portsmouth. This has resulted in the Portsmouth centre having

---

5   The phrase 'rough sleepers' collectively refers to homeless individuals who sleep outside, generally on the streets. It seeks to distinguish their circumstances from those accommodated in temporary shelters or by friends with homes. The lives of individual homeless people may, of course, cross between these categories.

higher quality premises but, at the same time, the Southampton sites are more closely connected to their communities and perhaps less intimidating for potential users. Finally, co-location at Portsmouth with an NHS treatment centre has, to an extent, obscured the PHC mission of the WiC in favour of an emphasis on minor surgery. Indeed, in terms of its website information, it is hard to find reference to the WiC aspect of the new development.

Research to date in fact suggests that the demographic profile of WiC visitors has relatively little overlap with genuinely disadvantaged groups. In a comparative study of WiC and GP consultations in England ($n = 6229$), it was found that '[w]alk-in centre visitors were more likely to be owner-occupiers (55 percent versus 49 percent; P<0.001), to have further education (25 percent versus 19 percent; P = 0.006), and to be white (88 percent versus 84 percent; P<0.001) than general practice visitors' (Salisbury et al. 2002a, 554). Men in the 25–44 year old age group were especially well represented (see Figure 14.5). The main reasons for attending a WiC were speed of access and convenience, and the most common health needs were viral illnesses, emergency contraception, minor injuries and wound dressings. The study concluded that 'NHS walk-in centres improve access to care, but not necessarily for people with the greatest health needs' (Salisbury et al. 2002a, 554).

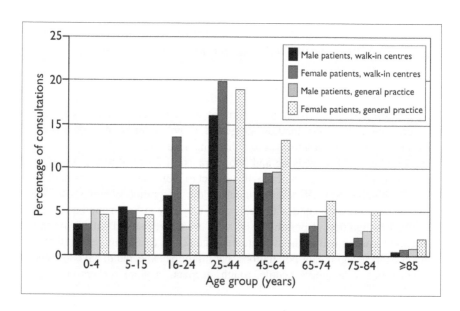

**Figure 14.5 Demographic profile for NHS walk-in centre and general practice consultations**

*Source*: Redrawn from Salisbury et al. (2002b).

*Health Care Provision for Homeless People*

We now turn to our second form of PHC service: non-statutory initiatives for homeless people. Southampton PCT inherited a long-standing commitment to health care for the homeless from its predecessor organizations (SPCT 2005). Building on the work of a charitable organization, the St. Dismas Society, a multi-agency project involving social care and health services was developed in the early 1990s. This involved part-time input from two GPs, working alongside general and psychiatric nurses. A central location that was known and used by homeless people was deliberately chosen. By 1993, the initiative had been transformed from a fixed-term project to an established service. Subsequent expansion saw the recruitment of social workers, increased psychiatric support staffing and a specialist in the health care of Roma, travellers and asylum seekers. The needs of homeless women were also recognized, with a deliberate decision to employ a female community psychiatric nurse.

In the late 1990s a decision was taken to reconstitute the project as a 'personal medical services' general practice focused on the health needs of homeless people. The majority of staff, including the GPs, are consequently now salaried PCT employees. Staff numbers have continued to grow and at the time of writing the team consisted of three GPs, four nursing staff, three mental health workers, three health visitors (specializing in care for children and families) and a social worker, as well as administrative support staff.

The Southampton Homeless Healthcare Practice is significant in that it registers patients without a permanent address. This ensures its relevance to its target population. People without a GP are a specific target, as are 'people who are homeless – those without security of tenure, including bedsits, bed and breakfast, hostels, refuges etc – as well as those sleeping rough' (SHHCP 2007). Consultations are available without appointments, as the practice uses a drop-in system to accommodate patients as they arrive. This approach does, however, have the disadvantage that people sometimes have to wait for care. A further constraint is that service hours are limited to 9:00am to 4:30pm during week days, with GPs only available on four mornings. Outside of these hours, potential users would need to attend the central WiC.

The Bodywise project run at the Portsmouth Foyer is an example of a more specialized non-statutory service. The term foyer here refers to residential accommodation facilities for young people who are either homeless, refugees, asylum seekers or on emergency placements (Foyer Federation 2007). Originating in projects that supported young workers in French cities, foyers provide a safe environment in which residents can access support that will help them towards independent living. As part of this mix of services, the Bodywise project offers health-related support. Its objectives are the enhancement of health knowledge and awareness among residents and support workers. It is supported financially by Portsmouth PCT and the Gatsby Charitable Trust, although this is currently a renewable rather than long-term arrangement.

In its first year of operation, Bodywise reported 37 residents as claiming lifestyle changes and 38 staff reporting improved health knowledge. The project carried out 256 individual health drop-in sessions for residents and 240 health-promoting activities or meetings, with 25 staff attending health training sessions (ECLN 2005). Activities covered drug awareness (including alcohol and tobacco), self-harm, contraception, parenting and sexual health, and ensuring that residents were up to date with immunizations (many were found to have missed standard early-youth immunization against tuberculosis). A breakfast club was developed to foster healthy eating. Running through all activities was an emphasis on informality, building trust, empowerment and encouraging participants to make contact with more mainstream health services.

In contrast to the health care treatment and delivery service offered by the Southampton Homeless Healthcare Practice, Bodywise clearly focuses on health promotion. Nonetheless, as a representative of the local PCT indicated: '[It] is a key project in terms of health. It targets a high risk group who tend to be socially excluded and helps to increase health access for those young people' (ECLN 2005). Without the backing of permanent funding, however, it remains an important but vulnerable initiative. Taken together, the two homeless health initiatives that we have considered exemplify different approaches and demonstrate how local circumstances and historic commitments underpin much of the development of PHC provision for socially excluded groups.

In both Southampton and Portsmouth, these facilities exist within a wider landscape of voluntary sector services for homeless people. A variety of agencies offer material support, including food, clothing and temporary accommodation, as well as advice and advocacy services. Some of these organizations also appear to function as informally therapeutic environments, by virtue of the physical safety, relational contact and consistency of welcome they offer. This is not to ignore the tensions and processes of exclusion that may occur within such settings (Parr 2000; Sager and Stephens 2003), but for some homeless people they evidently become supportive environments in their everyday negotiations of the city (Conradson 2003; Johnsen et al. 2005; Knowles 2000). Drawing on Buber's distinction between de-personalizing I-It relationships and respectful I-Thou relationships, Wen et al. (2007) argue that positive encounters in such places tend to be characterized by inclusivity, a lack of stereotyping and the absence of discrimination. This is a point reinforced by Ensign (2004), whose work in both street and clinical settings in the American city of Seattle highlighted the sensitivity of homeless youth to the qualitative dimensions of health care services.

In considering the positive potential of these project based PHC services, we should note that the examples considered here all require individuals to physically present themselves at the facility in question. To our knowledge, there are no genuinely mobile PHC services for disadvantaged groups in Portsmouth or Southampton, in the sense of nurses or doctors who physically travel to where people are on the streets. There are a number of mobile welfare services, such as local soup runs, which offer a more general mix of support that includes food, advice and clothing. For more explicitly health-related needs, however, it seems that local homeless people must currently present themselves at an existing NHS or NHS-funded facility. This is arguably a limitation of current PHC services with regard to meeting the needs of the most vulnerable and poor. The well recognized spatial unevenness in voluntary sector provision of health and welfare services is also an issue; in the absence of central planning, the creation of resilient non-statutory services is closely dependent upon local initiative and stable funding (Milligan and Conradson 2006).

**Conclusion**

We have examined two forms of PHC service, each of which represents a move beyond the traditional gateways of the GP and the A&E unit. The first, the NHS WiC, might be assumed to be a relatively effective method of increasing access to disadvantaged populations, particularly those whose way of life is significantly street-based (such as homeless people and some sex workers). However even in Southampton, where some of the existing WiCs are located in or near relatively deprived areas, the research did not suggest a significant increase in access from individuals within these groups. Indeed, more extensive studies find that the socio-economic profile of WiC visitors is often more affluent than that of general practice patients (Salisbury et al. 2002a; Salisbury 2007). WiCs have enhanced access to PHC, but apparently not to those groups whose health needs are the greatest. In terms of the geography, their location is also not unexpectedly a reflection of historical NHS land holdings rather than need. WiCs undoubtedly provide more localized PHC but they remain medicalized places, arguably more attractive to the affluent and 'busy' than the excluded.

Our second type of facility, the NHS-funded project based organization, appears more effective in addressing the health care needs of the most disadvantaged groups. In relaxing the registration requirement normally associated with access to general practice, organizations such as the Southampton Homeless Healthcare Practice certainly overcome what is an important first-order barrier for many homeless people. Over time, homeless people in the cities of Portsmouth and Southampton have come to recognize this and other voluntary sector organizations as sites where non-stigmatizing health care can be reliably accessed. Such agencies are generally well located, with services that are temporally accessible to the socially marginalized. That said, there remain vulnerable individuals who seldom engage with even these facilities. The health care needs of these groups are perhaps better addressed by genuinely mobile services, where health care professionals leave their organizational bases and seek to assist people on the street, in resolutely non-clinical environs. But the relatively dispersed and low-intensity nature of such assistance does tend to render it more expensive than either general practice or WiC care.

If primary care in the UK is to achieve the levels of accessibility and inclusivity to which the NHS aspires, and to which a genuinely comprehensive primary care system addressing the needs of marginal, excluded and disadvantaged populations points, then further action is clearly needed. Looking beyond medicalized health care towards the root social causes of much ill health among such populations suggests a need to move on from PHC, however well intentioned, to a broader primary care approach. Here health and social care would be integrated alongside services concerned with other welfare needs. It would be delivered in flexible ways and in partnership with client communities, rather than being framed by governmental agendas or constrained by the serendipities of local voluntarism. In the short-term, however, increased statutory investment in mobile PHC for the most disadvantaged groups is certainly an area for further consideration.

## Acknowledgements

We are grateful to the editors for their comments on an earlier version of this chapter, and to Lyn Ertl and Bob Smith in the Cartographic Unit at the University of Southampton for drawing the map and the graph.

## References

Adamson, J., Ben-Shlomo, Y., Chaturvedi, N. and Donovan, J. (2003), 'Ethnicity, Socio-economic Position and Gender – Do They Affect Reported Health-care Seeking Behaviour?', *Social Science & Medicine* 57: 895–904.

Blumenthal, D., Mort, E. and Edwards, J. (1995), 'The Efficacy of Primary Care for Vulnerable Population Groups', *Health Services Research* 30:1, 253–73.

Burt, J., Hooper, R. and Jessop, L. (2003), 'The Relationship Between Use of NHS Direct and Deprivation in Southeast London: An Ecological Analysis', *Journal of Public Health* 25, 174–6.

Chapman, J., Zechel, A., Carter, Y. and Abbott, S. (2004), 'Systematic Review of Recent Innovations in Service Provision to Improve Access to Primary Care', *British Journal of General Practice* 54:502, 374–381.

Conradson, D. (2003), 'Spaces of Care in the City: The Place of a Community Drop-in Centre', *Social and Cultural Geography* 4:4, 507–25.

Daiski, I. (2007), 'Perspectives of Homeless People on their Health and Health Needs Priorities', *Journal of Advanced Nursing* 58:3, 273–81.

DoH (Department of Health) (2000), *Pharmacy in the Future: Implementing the NHS Plan* (London: DoH).

Ensign, J. (2004), 'Quality of Health Care: The Views of Homeless Youth', *Health Services Research* 39:4, 695–708.

Feldman, R. (2006), 'Primary Health Care for Refugees and Asylum Seekers: A Review of the Literature and Framework for Services', *Public Health* 120, 809–16.

Gatrell, A. (2002), *Geographies of Health: An Introduction* (Oxford: Blackwell).

Gulliford, M. and Morgan, M. (2003), *Access to Health Care* (London: Routledge).

HCHC (House of Commons Health Committee) (2004), *GP Out-of-Hours Services* (London: The Stationery Office).

Hills, J. and Stewart, K. (2005), *A More Equal Society? New Labour, Poverty, Inequality and Exclusion* (Bristol: Policy Press).

Johnsen, S., Cloke, P. and May, J. (2005a), 'Day Centres for Homeless People: Spaces of Care or Fear?', *Social and Cultural Geography* 6:6, 787–811.

Johnsen, S., Cloke, P. and May, J. (2005b), 'Transitory Spaces of Care: Serving Homeless People on the Street', *Health & Place* 11:4, 323–36.

Joseph, A.E. and Phillips, D.R. (1984), *Accessibility and Utilization: Geographical Perspectives on Health Care Delivery* (New York: Harper & Row).

Knowles, C. (2000), *Bedlam on the Streets* (London: Routledge).

LHPC (London Health Planning Consortium) (1981), *Primary Health Care in Inner London* (London: Department of Health and Social Security, Acheson report).

Majeed, F.A., Chaturvedi, N., Reading, R. and Ben-Schlomo, Y. (1994), 'Equity in the NHS: Monitoring and Promoting Equity in Primary and Secondary Care', *British Medical Journal* 308, 1426–29.

McColl, K., Pickworth, S. and Raymond, I. (2006), 'Project: London – Supporting Vulnerable Populations', *British Medical Journal* 332:7533, 115–7.

Milligan, C. and Conradson, D. (eds) (2006), *Landscapes of Voluntarism: New Spaces of Health, Welfare and Governance* (Bristol: Policy Press).

Moon, G. and North, N. (2000), *Policy and Place: General Practice in the UK* (London: Macmillan).

Morris, S., Sutton, M. and Gravelle, H. (2005), 'Inequity and Inequality in the Use of Health Care in England: An Empirical Investigation', *Social Science & Medicine* 60, 1251–66.

Pantasiz, C., Gordon, D. and Levitas, R. (2006), *Poverty and Social Exclusion in Britain: The Millennium Survey* (Bristol: Policy Press).

Parr, H. (2000), 'Interpreting the "Hidden Social Geographies" of Mental Health: Inclusion and Exclusion in Semi-institutional Places', *Health & Place* 6, 225–37.

Pope, C., Chalder, M., Moore, L. and Salisbury, C. (2005), 'What Do Other Local Providers think of NHS Walk-in Centres? Results of a Postal Survey', *Public Health* 119, 39–44.

Quilgars, D. and Pleace, N. (2003), *Delivering Health Care to Homeless People: An Effectiveness Review* (York: University of York).

Sager, R. and Stephens, L.S. (2005), 'Serving Up Sermons: Clients' Reactions to Religious Elements at Congregation-run Feeding Establishments', *Nonprofit and Voluntary Sector Quarterly* 34:3, 297–315.

Salisbury, C. (2007), 'NHS Walk-in Centres', *British Medical Journal* 334, 808–9.

Salisbury, C., Manku-Scott, T. and Moore, L. et al. (2002a), 'Questionnaire Survey of Users of NHS Walk-in Centres: Observational Study', *British Journal of General Practice* 52:480, 554–560.

Salisbury, C., Chalder, M. and Manku-Scott, T. et al. (2002b), 'What is the Role of Walk-in Centres in the NHS?', *British Medical Journal* 324, 399–402.

Schanzer, B., Dominguez, B., Shrout, P.E. and Caton, C.L.M. (2007), 'Homelessness, Health Status, and Health Care Use', *American Journal of Public Health* 97:3, 464–9.

Shaw, M., Dorling, D. and Brimblecombe, N. (1999), 'Life Chances in Britain by Housing Wealth and for the Homeless and Vulnerably Housed', *Environment and Planning A* 31:12, 2239–48.

Smaje, C. and Le Grand, J. (1997), 'Ethnicity, Equity and the Use of Health Services in the British NHS', *Social Science & Medicine* 45, 485–96.

Townsend, P. and Davidson, N. (1982), *Inequalities in Health* (Harmondsworth: Penguin).

Townsend, P., Davidson, N. and Whitehead, M. (1992), *Inequalities in Health: The Black Report and the Health Divide* (Harmondsworth: Penguin).

Wen, C.K., Hudak, P.L. and Hwang, S.W. (2007), 'Homeless People's Perceptions of Welcomeness and Unwelcomeness in Health Care Encounters', *Journal of General Internal Medicine* 22:7, 1011–7.

**Internet-based References**

Dixon, A., Le Grand, J. and Henderson, J. et al. (2003), *Is the NHS Equitable? A Review of the Evidence* (London: London School of Economics) Discussion Paper 11, http://www.swpho.nhs.uk/resource/item.aspx?RID=14431, accessed September 15, 2007.

ECLN (Engaging Communities Learning Network) (2005), *Stories that Can Change Your Life: Communities Challenging Health Inequalities* (Winchester: ECLN), http://www.natpact.nhs.uk/uploads/2005_Jan/Health_Inequalities.pdf, accessed September 15, 2007.

Foyer Federation (2007), http://www.foyer.net/mpn/topic.php?topic=3, accessed October 12, 2007.

Mercury Health (2007a), *Mercury Health Website*, www.mercuryhealth.co.uk, accessed September 12, 2007.

Mercury Health (2007b), *Mercury Health Website*, http://www.mercuryhealth.co.uk/index.php?id=42, accessed October 15, 2007.

Office of the Deputy Prime Minister (2004), *The English Indices of Deprivation 2004* (revised) (London: Office of Deputy Prime Minister), http://www.communities.gov.uk/documents/communities/pdf/131209, accessed November 15, 2007.

SHHCP (Southampton Homeless Healthcare Practice) (2007), http://www.homelesshealthcare-southampton.nhs.uk/welcome/homelesshealthcare, accessed September 12, 2007.

SPCT (Southampton Primary Care Trust) (2005), *History of the Homeless Healthcare Team*, http://www.homelesshealthcaresouthampton.nhs.uk/welcome/homelesshealthcare/historyofthehht1, accessed September 12, 2007.

SPCT (Southampton Primary Care Trust) (2007), *Press Release*, http://www.southamptonhealth.nhs.uk/trust/news/citys-third-nhs-walk-in-centre, accessed September 12, 2007.

Walker, P. (2007), 'NHS Spending to Rise 4% to £110bn', October 9, 2007, *The Guardian*, http://www.guardian.co.uk/society/2007/oct/09/publicservices, accessed December 10, 2007.

# PART 4
# Agenda Setting

Chapter 15

# The Geographies of Primary Health Care: A Summary and Agenda

Valorie A. Crooks and Gavin J. Andrews

In this, the concluding chapter, we aim to both summarize what has been written about in the previous contributed chapters and also put forward an agenda for further geographical research on primary health care (PHC) both by health geographers specifically and more broadly by health science, health professional, and social science disciplines interested in geographic phenomena. We start by revisiting three themes integral to most definitions of PHC, including the foundational one set out in the Alma-Ata Declaration, that were introduced in Chapter 1, specifically: access, equity, and community. We then move to identify four emergent themes from the preceding contributed chapters: continuum and continuity of care, care sites, technology, and transforming PHC. These themes are crosscutting and certainly contribute to establishing and agenda for furthering the geographies of PHC.

In Chapter 1 we stated that through producing this book we aim to interrogate 'the landscapes that inform the very nature and principles of PHC and ultimately shape how this care is both delivered and received.' In the third subsection of this chapter we revisit this aim by articulating what we refer to as the 'landscapes of PHC.' Landscape concepts are important in the sub-discipline of health geography and so we view this as an important opportunity to synthesize this landscape before more fully articulating a research agenda. In the fourth and final subsection we do just this; we revisit ideas put forth in Chapters 1 and 2, reflect on the contributed chapters, and also on the syntheses written in this chapter in doing so. It is our hope that this agenda speaks to geographers and non-geographers alike who are interested in exploring the geographies of PHC.

## Revisiting Access, Equity, and Community

In Chapters 1 and 2 we identified access, equity, and community to be PHC themes that resonated most closely with long-standing interests of health geographers and their research regarding health care. It is no surprise, then, that these themes have come up repeatedly throughout the book. In this subsection we revisit these themes and summarize what we have learned about them throughout the contributed chapters. Of course, it goes without saying that our summary is only partial given that very much can be gleaned from the previous twelve chapters about these themes and how they relate to PHC. We also recognize that summarizing these themes separately is somewhat of an abstraction from how they have been discussed throughout the book as they are very much interrelated in relation to PHC and one typically evokes discussion of the others. Moreover, other crosscutting elements of PHC more broadly such as workers and work and service categories such as clinics and homes are discussed within and between these themes.

*Access*

As the preceding chapters have clearly shown, there is no one single way to understand or conceptualize access as it relates to PHC. Though providing accessible care is clearly a hallmark of the PHC approach envisioned in the Alma-Ata Declaration (see Chapter 1), how notions of access have been put into practice and place differ significantly across jurisdictions. Importantly, as Neil Hanlon points out in Chapter 3, having access to PHC means much more than being physically proximal to sites of care. He contends that we must factor utilization into considerations of access while Nadine Schuurman, in Chapter 4, adds that vulnerability as it relates to socio-economic status and geographic location is an important consideration, and David Conradson and Graham Moon, in Chapter 14, contend that capacity must also be considered (i.e., does the clinic have the ability to take on new patients?). Though geographic access alone is not a nuanced enough understanding, Nadine Shuurman along with Robin Kearns and Pat Neuwelt in Chapter 12 suggest that the physical landscape as determined by topography and other features is an important barrier to people's access to primary care and PHC (particularly in very mountainous or rural areas) that must be considered in service siting and delivery. Finally, desired access (i.e., what kind of care do people want?) was raised by Daniel Hollenberg and Ivy Bourgeault in Chapter 10 to be an important factor in providing PHC.

Ideas of access were invoked in many interesting ways by contributing authors. This includes that nurses and other PHC providers can enhance patients' system navigation through the development of relationships that facilitate information-sharing and other such guidance (Jennifer Lapum et al. Chapter 8). In Chapter 7, Gina Agarwal notes that PHC providers can gain access to patients' lifeworlds and routines through providing care in the home. Meanwhile, while also considering the home, Nicole Yantzi and Mark Skinner, in Chapter 13, contend that having access to support, information, and resources is essential for informal providers' abilities to provide care in the home.

Factors posing as constraints on access featured prominently in several chapters, including that: a person's reading of the clinic environment may deter him or her from accessing care from a specific site (Crooks and Agarwal, Chapter 11); siting services in areas of deprivation, thereby enhancing geographic accessibility, does not ensure use by those in most need (David Conradson and Graham Moon); health human resource availability in part determines access to care (Janine Wiles and Mark Rosenberg, Chapter 5); and people's socio-economic circumstances and decision-making may necessitate having to not seek out care regardless of its accessibility (Leah Gold, Chapter 6; Ross Barnett and Pauline Barnett, Chapter 9).

*Equity*

Chapter 1 introduced that equity was at the centre of the World Health Organization's (WHO) foundational vision of PHC and has been referenced in numerous definitions of such care ever since. Janine Wiles and Mark Rosenberg conceptualize equity in PHC as care that prioritizes those most in need. Given this, it is not surprising that many contributors related equity and access in their discussions. David Conradson and Graham Moon, for example, point out that equity in access to care has been a longstanding goal of the British government. Meanwhile Jennifer Lapum and colleagues' review of the historical origins of PHC nursing reveals that nursing has always attempted to address inequities in access to care through practice since its inception. Nadine Schuurman, however, reminds us that equity as it relates to access to PHC must be carefully considered, particularly when looking at how it is played out at the intersection of multiple axes of difference (e.g., an individual experiencing rural vulnerability and of low socio-economic status). Furthering the conceptualization of equity, Neil Hanlon notes that it is 'distributional equity' that is most evoked in relation to PHC because of its emphasis on first-contact and continuing primary care (versus specialist care). Interestingly, an important point about the very nature of equity in PHC was made by Leah Gold who, in discussing a Peruvian case study, contends that selective PHC (discussed in Chapter 1) prioritizes efficiency over equity.

Not surprisingly, issues of equity as it relates to PHC came up repeatedly in discussions of vulnerable and marginalized groups. Inequities in access to and delivery of PHC and/or primary care for indigenous groups featured prominently in Leah Gold, Robin Kearns and Pat Neuwelt, and Ross Barnett and Pauline Barnett's discussions. Their chapters reveal the importance of addressing unequal access to care across groups through using strategies to promote equity in service provision and health outcomes. A particular barrier to doing this noted in these chapters is the presence of user fees which serve as significant barriers to care for marginalized and vulnerable groups. Another challenge related to equity in PHC was raised by Valorie Crooks and Gina Agarwal. Drawing on interviews with depressed women, they show that these patients are very aware of the clinic environment and suggest that this may be due to an increased sensitivity to their surroundings as a result of experiencing depression. In considering the larger implications of this finding and those from other studies, they ask: how do we determine which needs to both accommodate and prioritize? This question is important because different clinic users have different 'readings' of the clinic space and perhaps differing needs. This can be extended more broadly to other issues of equity where there may be multiple, and even competing, needs across groups that must somehow be ranked and/or evaluated in terms of priority.

Equity is an issue of importance not only to care recipients but also caregivers. Nicole Yantzi and Mark Skinner, for example, contend that equity in PHC not only relates to the care that is delivered to individuals but also to the differential valuing of those who give it. They make this point through their larger contention that informal providers of PHC in the home are undervalued and their role as key givers of PHC is mostly unrecognized. The implications of this are spelled out in detail in their chapter. Gina Agarwal reflects on her own practice as a general practitioner (GP) and in doing so notes that patients with higher socio-economic status expect more from her as a clinician – specifically a better quality of care – the outcome of which is that she must actively consider how she equitably provides care across groups with differing levels of education and access to information that ultimately create different expectations. Her point serves as a reminder that we must not only consider equity in access to PHC but also equity in practice of PHC.

*Community*

While reference to community is a defining feature of PHC, putting the idea of providing care by, for, and in the community that seeks the input of community members into practice is challenging. This point is at the centre of Robin Kearns and Pat Neuwelt's chapter whereby they explore what community participation means in relation to primary care provision in New Zealand. An important point they raise is that effective community involvement necessitates getting all those who have a stake in a particular service involved, where stakeholders extend well beyond patients and service providers. Because of the limited ways in which community is interpreted in relation to PHC, Leah Gold expresses criticism toward the World Bank's contention that user fees are part of community participation in PHC. As noted above, such fees can pose as a significant barrier to care for economically marginalized individuals in particular.

Community as it relates to service provision and providers was an identifiable theme across several of the chapters. Lapum et al. for example, note that the development of the nurse-client relationship enables nurses to gain a greater understanding of the communities they practice in. This comment is echoed by Gina Agarwal as it relates to GPs, who also observes that these providers have a professional responsibility to the community. Daniel Hollenberg and Ivy Bourgeault's discussion of integrative medicine serves as a reminder that communities of providers and users very much shape this type of PHC provision. Indeed, where scepticism has often occurred both in the general population and conventional medical community, providers and users of complementary and alternative medicine (CAM) often identify themselves as a cohesive counter movement, and even culture. David Conradson and Graham Moon further point out that community need may in part dictate hiring practices. Finally, Nicole Yantzi and Mark Skinner note that care delivered in community settings is becoming increasingly complex and sophisticated and this has significant implications for informal human health resource personnel (i.e., informal caregivers) in particular.

From the above discussion it can be understood that 'community' was interpreted quite differently across the chapters; meanwhile these interpretations are united in that they were all made in relation to PHC. As Janine Wiles and Mark Rosenberg contend, community is not only a site but is also an important scale of PHC provision and consumption. Ideas of community were evoked at many scales by the contributors. At a macro-scale, Ross Barnett and Pauline Barnett point out that health care reform has significant outcomes for communities and professional groups alike. Meanwhile, Neil Hanlon notes that investment in PHC – something likely done at a macro-scale – is reflective of a commitment to community development. At perhaps the most micro-scale discussed, Valorie Crooks and Gina Agarwal question the role the community should play in determining the clinic environment. Because PHC is to be delivered in communities, and for communities, they wonder if and how communities should determine what clinics look like, including their aesthetics and layouts.

## Emerging and Crosscutting Themes for Further Exploration

Chapters contributed to this book were organized into three subsections relevant to the geographies of PHC: (1) practice and delivery; (2) people; and (3) places and settings. Further, as we discussed in the previous section, access, equity, and community serve as important themes related to both PHC and also geographic research on health care more broadly. While these subsections and themes provide some useful organizational structure both to the book and to thinking about the geographies of PHC, it is not surprising that important crosscutting themes that warrant further exploration have emerged from the contributed chapters. In this subsection we identify four such themes: (1) continuum and continuity of care; (2) care sites; (3) technology; and (4) transforming PHC.

### Continuum and Continuity of Care

Establishing continuity of care (i.e., a relationship of trust between doctor and patient [interpersonal/relational continuity], having a regular site of care [longitudinal continuity], and having information availability within and between sites [informational continuity]) and a continuum of care (i.e., the integration of care across settings and providers) for PHC provided both within and beyond formal primary care systems are not only important but are also indicators of care quality. Continuity of care, for example, is positively correlated with having access to health care and improved patient satisfaction (Saultz and Albedaiwi 2004). Janine Wiles and Mark Rosenberg further point out that continuity is a defining characteristic of PHC. Meanwhile, Robin Kearns and Pat Neuwelt contend that the continuum of PHC must be considered not only as it relates to care but also as it relates to community (i.e., the continuum of participation from an individual through to an entire community).

Not surprisingly, Nicole Yantzi and Mark Skinner note that in Canada an aim of PHC reform is aiming to strengthen both the continuity and continuum of care. They further contend that providers of PHC in the home are undervalued in the role they play in the continuum of care and that they are essential in establishing continuity of care within and beyond PHC settings. They suggest that this is an area such reform should address. Neil Hanlon notes that roster sharing is also another element of PHC reform that has the capacity to enhance continuity of care, particularly when GPs leave a practice.

In their discussion of integrative medicine, Daniel Hollenberg and Ivy Bourgeault contend that integrative clinics, as a new space of PHC provision, enable more coordinated care across the continuum and enhance continuity through the provision of multiple services (operating across professional philosophies of care) from one site. Such a contention directly implicates the role of space. Gina Agarwal's discussion of continuity of care also does this in that she discusses how providing care over time to patients allows GPs to develop a greater understanding of their 'family geographies.' Interestingly, this 'extended' understanding of patients' lives is established through interpersonal interactions undertaken within the clinic.

*Care Sites*

Attention to the more spatial aspects of discussions in the preceding chapters has, not surprisingly, led to our identification of 'care sites' as an important crosscutting theme. Sites of care and their interrelationships are an important feature of PHC because, as Valorie Crooks and Gina Agarwal note, 'care is always delivered *somewhere* (i.e., in or from a particular place).' Leah Gold, for example, points out that the clinic may not always be viewed as the best site of care, as is the case in rural Perú where care in the home may be prioritized. Robin Kearns and Pat Neuwelt, meanwhile, remind us that sites of care can have multiple roles in communities and can serve as places of gathering and information exchange in addition to health care provision. They, along with Valorie Crooks and Gina Agarwal, suggest that care sites have the potential to promote health. Because of all these factors, we must be attentive to care sites in relation to transforming PHC and specifically considerations of equity in terms of sites of care and not just care distribution and quality.

Discussion of specific care sites featured prominently throughout the chapters. Nadine Schuurman, for example, focused on the hospital and notes that they are important sites of PHC provision. Gina Agarwal suggests that the provision of care in the home by GPs is beneficial to both parties but that the boundaries of privacy when entering the private space of a patient's home must be carefully negotiated. Further, both she and David Conradson and Graham Moon suggest that there are benefits to be had by at risk, vulnerable, and more transient populations from care that offers more geographic flexibility through the provision of mobile care.

*Technology*

Technologies have multiple uses and applications in health care. Central to the Declaration of Alma-Ata put forth by the WHO (see Chapter 1) is the recognition that technologies should inform PHC practice and provision. Such technologies can include those related to medical/health informatics, such as electronic medical records, those related to the Internet, including the practice of 'cybermedicine' and the use of Web 2.0 to facilitate the creation of virtual communities specific to health issues, and those which enable the provision of care, such as diagnostic machinery, among others. Furthermore, and best exemplified by Nadine Schuurman's chapter, the use of technologies in research (e.g., GIS) can also further PHC inquiry and expand our knowledge about such care.

Much was said about technology and its relationship to PHC throughout the contributed chapters. It was frequently noted that technology is facilitating access to care in ways that were not previously possible, including to the practice of PHC by nurses (Jennifer Lapum et al.) and GPs (Gina Agarwal). There were, however, numerous critiques of technology. Valorie Crooks and Gina Agarwal, for example, point out that the use of technologies in practice may create literal and/or perceived distance between provider and patient thereby negatively affecting the establishment of a therapeutic relationship between these parties. Neil Hanlon remarks that in spite of technological advancements, the preferred mode of PHC provision remains face-to-face. In her case study of PHC clinics in Perú, Leah Gold demonstrates that while technology is an important indicator of care quality, low-tech interventions still have effectiveness with accessing hard-to-reach populations (e.g., health promotion posters placed in clinics and elsewhere that convey basic health information). Nicole Yantzi and Mark Skinner contend that technology blurs boundaries between care sites and also between formal and informal providers. Janine Wiles and Mark Rosenberg argue that the purchase and use of technologies in PHC practice is shaped by those who have decision-making power. Finally, David Conradson and Graham Moon note that having access to technologies (e.g., Internet, phones) cannot be prerequisites for individuals' care receipt as those in most need of care are often least likely to have access to them.

*Transforming Primary Health Care*

The final significant crosscutting theme we have identified is discussion related to transforming PHC. While it served as a significant point of discussion in some chapters, in many others it was reflected on in the concluding section where implications of the points made in the chapter were drawn out. Gina Agarwal, for example, points out that transforming PHC must involve considering the practice-based geographies of different provider groups, including GPs, in her concluding discussion. Similarly, Jennifer Lapum et al. reflect on the key role that nurses can and should play in transforming PHC including through directing its leadership at multiple scales, local through global, in their conclusion. They more broadly note that notions of biomedical dominance must be overcome in order to transform PHC, a point echoed in Daniel Hollenberg and Ivy Bourgeault's chapter.

In their chapter on the role of scale in PHC, Janine Wiles and Mark Rosenberg suggest that we must be attentive to the fact that transformations at one scale have implications for PHC at others. Ross Barnett and Pauline Barnett concur, demonstrating both that primary care transformation happens quite differently across nations and that there are important connections between governing parties and strategies adopted for primary care transformation. Their discussion centres around primary care systems, which Neil Hanlon contends is an important area of transformation as part of PHC reform initiatives.

## Articulating the Landscapes of Primary Health Care

The term 'landscape' is used in many ways throughout the larger discipline of geography, including both within and across human and physical geography. As Gesler and Kearns (2002) note, the term also has popularized meanings, including in relation to the design and aesthetic of exterior spaces (e.g., landscape design, landscape architecture). Because of this, here we first introduce 'landscape' and landscape concepts as they are used by human geographers, and health geographers specifically, prior to articulating the what we refer to as the 'landscapes of PHC'.

In its broadest sense, geographers use the term landscape to refer to 'the appearance of an area, the assemblage of objects used to produce that appearance, and the area itself' (Duncan 2000, 429). Clearly this is a broad interpretation. More relevant to health geography is that landscapes both shape social, cultural, and political structures, systems and practices and are simultaneously shaped by them. Further to this, landscapes typically characterize interrelationships between and variances across people and environments, systems, and particular places. As Williams (1999) points out, landscapes are also ever changing. Curtis (2004) furthermore observes that there are complex interrelationships between landscapes. Gesler and Kearns (2002) contend that people interpret landscapes in different ways. These are all factors which add to the complexity of articulating a particular landscape.

In health geography, landscape concepts – usefully referred to as 'conceptual landscapes' by Curtis (2004, 23) – are used to characterize the processes mentioned in the previous paragraph and how, why, and where they produce variation as it relates to both health and health care. Different landscape concepts offer distinct explanatory perspectives – both applied and theoretical – on this variation, including its causes and outcomes. A number of landscape concepts are frequently drawn upon by health geographers, including: therapeutic landscapes (place-based factors promoting healing in places and places that heal – see Williams 1999,

2007); landscapes of voluntarism (factors contributing to the roles of the voluntary sector and informal carers in care provision – see Milligan and Conradson 2006); landscapes of healing (factors that ultimately shape whether and how healing is experienced – see Gesler and Kearns 2002); and the landscape of health inequality (those factors that contribute to differences in health outcomes across space – see Curtis 2004). These latter two concepts in particular are meta-landscapes in that they are characterized by the interrelationships between multiple landscape concepts. For example, Curtis (2004) argues that the landscape of health inequality is made up of five landscapes: (1) therapeutic landscapes; (2) landscapes of power and resistance; (3) landscapes of poverty and wealth; (4) landscapes of consumption; and (5) ecological landscapes. Gesler and Kearns (2002), meanwhile, suggest that both the landscapes of consumption and therapeutic landscapes, among others, are implicated in the larger landscapes of healing.

Based on the importance of PHC, its relationship to core geographic concepts (e.g., access, equity, community), and recognition that such care is informed by complex and interrelated factors, we suggest that the 'landscapes of PHC' is an important conceptual construct worthy of further exploration and fuller articulation. This landscape can be characterized by the people, practices, and places involved in giving and receiving such care. Based on what has been shared in the preceding chapters, this landscape is shaped by structures such as the political organization of societies that in part determines infrastructure for such care and the desire and ability to achieve better health outcomes, systems such as primary care systems that deliver much first-contact medical care and welfare systems that enable and constrain access to needed health and social care for marginalized groups in particular, and the very practices of health and social care more broadly, including those of specific professional groups and informal caregivers, along with the cultural practices of patients. These structures, systems, and practices weave together to create the landscapes of PHC.

Gesler and Kearns (2002) point out that language is an important part of any landscape. This is certainly true for the landscapes of PHC. In Chapter 1 we discussed in detail how PHC has multiple interpretations (e.g., an approach versus a philosophy of care) and definitions within and between jurisdictions at multiple scales. These more scripted elements of PHC are essential components of this conceptual landscape in that they determine what PHC is, who this care is for, who gives such care, and how care is paid for, among other things. The language of PHC and its specific definitions have significant implications for multiple groups including both providers and patients.

Connectedness is another important aspect of the landscapes of PHC. Connections between care sites, services, and providers, for example, serve as important interconnections within and beyond caring environments. There are also important connections between how PHC is conceptualized and/or interpreted and how it is practised. Numerous other fundamental connections have been established throughout the preceding chapters. While connectedness between specific factors/variables of PHC is important, interconnections between the landscapes of PHC and other landscape concepts certainly are as well. Therapeutic landscapes are implicated in the landscapes of PHC through the establishment of potentially healing sites and spaces of care delivery and practice, the landscapes of voluntarism are related through the use of voluntary-sector and volunteer labour to provide PHC, and the landscapes of power and resistance are evoked through determinations of who 'community' is and how 'community participation' can and should be operationalized. These are but a few examples of the many ways in which the landscapes of PHC have a connectedness to other health and health care-related conceptual landscapes.

## Geographies of Primary Health Care: Setting the Agenda

Based on what has been discussed throughout this book and particularly in Chapter 1, we believe PHC can be best characterized (for the purposes of agenda-setting) as: health care and certain forms of social care given at the first point of contact – although such care may be continuing – both within and beyond the primary care system that addresses specific needs, promotes health, and assists in developing and maintaining wellbeing that is provided equitably and accessibly within the community for all its members using their input. Features we view to be hallmarks of this approach include references to equity and access in care provision as well as to care that is delivered by, for, and within communities for which the input of community members has been sought out. Important to this characterization is that PHC can involve both health and social care. This comes in recognition of the fact that factors beyond access to health care are important determinants of health and play key roles in promoting health and wellbeing which are goals of PHC. The provision of social care through programs that facilitate access to food, education, and shelter, for example, are central to the PHC philosophy and mandate as they allow for the achievement of health and wellbeing. Equally important is that PHC is not equated with primary care, nor is it care delivered exclusively within a formal primary care system. Based on this characterization, in this final subsection of the book we lay out an agenda for further exploration of the geographies of PHC.

Future geographical research agenda on PHC needs to engage with continuum and continuity of care issues across time and space. Here continuum and continuity should be considered in relation to two essential features, the first of which is service reforms. Research here should engage with the potential for PHC reform to enhance system connections (production) and patient journeys (consumption) within and between different forms of PHC and associated care practices. The second essential feature is service settings. Such research should consider system connections and patient journeys between different sites of PHC (categories such as homes, clinics and hospitals each require specific attention), to illuminate the true character and meanings of these places. Indeed, focused inquiry is required on what it means to provide PHC, particularly in terms of challenges and experiences, in and across specific places.

Throughout this book we have identified many themes that should be drawn upon in further establishing a research agenda. Earlier in the chapter, for example, it was noted that technology was a clear theme central to the geographies of PHC emergent from the discussions in the preceding chapters. There is a great need for further research about the roles of and uses for technologies in such care from a geographical perspective. First, and more generally, the role of technology in PHC requires consideration in terms of how it bridges care sites and communities such as through information linking and sharing between settings. Second, consideration must be given to how technologies impact upon places and people's lived experiences within them (e.g., how it may change the character of care received and social interactions in situ). Third, we must also investigate further how technology creates 'non-physical' PHC space itself. Here we are thinking of cyberspaces such as websites where physical co-presence does not occur, but where PHC is received or advised on.

This research agenda involves combining different perspectives and approaches from the health sciences and health geography, as outlined in Chapter 2. Whereas health geography has been successful at displaying broad macro-spatial features of health care, and more recently has identified the symbolism, identity, and attachments of health care places, the health sciences have demonstrated how geographies exist in the concerns and work activities of professionals and practitioners. Whilst both approaches are important, we would like to see a greater cross-fertilization of ideas and approaches. This would, for example, involve health scientists extending their explorations of human geography and undertaking some explicit considerations of how – beyond health geography – other fields of inquiry and sub-disciplines in human geography might be drawn on. Such fields might include, for example, emotional geography, children's geography, caring geography, geography of aging, and geographies of consumption. Meanwhile sub-disciplines might include, for example, rural geography, urban geography, social geography, economic geography, and population geography. What perspectives, theories, methods, and data can they each lend the research endeavour? These are the kinds of questions that need to be asked.

A greater cross-fertilization of ideas and approaches in the geographies of PHC would also involve health geographers moving away from twin streams of research on 'disease' and 'services' which artificially distance production from consumption. Geographers must attempt to study the (re)production of PHC in its fullest sense. In such an endeavour, post-medical thinking in the sub-discipline that involves assumptions that more applied research on health care is in some way uncritical or subservient and that research investigating the impacts or neglects of care on patients is more critical, sensitive, and needed should be avoided (see Andrews and Evans 2008). For geographers, part of a new approach to researching heath care needs to involve some explicit considerations of how other disciplines can assist in and inform our study. For example, whilst the sociology of health care organizations and work might provide critical theoretical insights in terms of access, health professional disciplines might help geographers to get closer to services, providers, and patients and critically assess the regulations, standards, policies, pressures, and groups that guide and affect them. Health professional literatures and debates might also provide theoretical inroads into important practice debates. Related to this point is that geographers clearly cannot hope to be the sole experts on PHC and as such should work in multidisciplinary teams and projects to address questions central to more geographical research agendas, as some have already started to do.

The chapters in this book demonstrate a solid foundation upon which to further develop the geographies of PHC as an important focus within the sub-discipline of health geography. The current and growing spatial awareness in all PHC areas research – reflected by the nurses, GPs, geographers, and sociologists who have contributed chapters to this book – directly acknowledges that PHC, perhaps more than any other form of health care, has spatial features that are fundamental to its existence and success. There are many varieties of health professions, occupations, and services that contribute to PHC. Every day, countless millions of workers and patients interact worldwide in the hundreds of thousands of places that exist for PHC. Given the volume and range of this global activity, academics would do well to further understand and articulate the geographies involved in the giving, receiving, organizing, and administering of PHC. Furthermore, in acknowledging that PHC is a global activity and is played out at scales from the global through to specific site of cares, consideration should usefully be given to how major global issues (and their local outcomes) such as climate change, aging populations, poverty and the growing income gap between individuals and nations, and emerging health crises, among others, have implications for PHC. The spatial dynamics of these issues make the geographic study of PHC ever more important.

In sum, numerous research questions and themes inform the furthering of the geographies of PHC as an important area of health geography inquiry. Engaging in such research will also allow for developing a fuller understanding of the landscapes of PHC that we outlined in the previous subsection. Certainly, areas of research inquiry and defining features of this conceptual landscape are sure to change over time as new health and health care issues emerge and take priority, as researchers from other disciplines further engage in exploring the geographies of PHC, and as geographers establish even greater connections with other disciplines and health professional groups engaged in PHC research and practice.

## References

Andrews, G.J. and Evans, J. (forthcoming), 'Understanding the Reproduction of Health Care: Towards Geographies in Health Care Work', *Progress in Human Geography*.

Curtis, S. (2004), *Health and Inequality* (London: Sage).

Duncan, J. (2000), 'Landscape', in R.J. Johnston, D. Gregory, G. Pratt and M. Watts (eds), *The Dictionary of Human Geography*, 4th edition (Oxford and Massachusetts: Blackwell Publishers Ltd.).

Gesler, W.M. and Kearns, R.A. (2002), *Culture/Place/Health* (London and New York: Routledge).

Milligan, C. and Conradson, D. (eds) (2006), *Landscapes of Voluntarism: New Spaces of Health, Welfare and Governance* (Bristol: Policy Press).

Saultz, J.W. and Albedaiwi, W (2004), 'Interpersonal Continuity of Care and Patient Satisfaction: A Critical Review', *Annals of Family Medicine* 2:5, 445–51.

Williams, A. (ed.) (1999), *Therapeutic Landscapes: The Dynamic Between Place and Wellness* (Lanham: University Press of America).

Williams, A. (ed.) (2007), *Therapeutic Landscapes* (Aldershot and Burlington: Ashgate).

# Index